The Human Face of Warfare

The Human Face of Warfare

Killing, Fear and Chaos in Battle

Edited by Michael Evans and Alan Ryan

ALLEN & UNWIN

Allen & Unwin
9 Atchison Street
St Leonards NSW 2065
Australia
Phone: (61 2) 8425 0100
Fax: (61 2) 9906 2218
Email: frontdesk@allen-unwin.com.au
Web: http://www.allenandunwin.com

National Library of Australia
Cataloguing-in-Publication entry:

The human face of warfare: killing, fear and chaos in battle.

  Bibligraphy.
  Includes index.
  ISBN 1 86508 374 7.

  1. War—Psychological aspects. 2. War victims—Mental
  health. 3. War neuroses. 4. Women and war. I. Evans,
  Michael, 1953-. II. Ryan, Alan, 1963-.

616.85212

Set in 10.5/12 pt Bembo by DOCUPRO, Sydney
Printed by CMO Image Printing Enterprise, Singapore

10 9 8 7 6 5 4 3 2 1

Contents

Tables and Figures

Abbreviations

AAR	after action review
ABC	Australian Broadcasting Commission
ADC	aide-de-camp
ADF	Australian Defence Force
AIF	Australian Imperial Force
BOS	battle operating system
C3I	command, control, communication and intelligence
CCF	critical combat function
CDA	Centre for Defence Analysis (United Kingdom)
CDF	Chief of the Defence Force (Australia)
CIS	Combat Information Support/System
CNN	Cable News Network
COMAST	Commander Australian Theatre (Australia)
CTC	combat training centre
CRMT	crew resource management training
DFSC	Defence Food Science Centre (Australia)
DPS	Defence Psychology Service (Australia)
DVA	Department of Veterans' Affairs (Australia)
DSTO	Defence Science and Technology Organisation (Australia)
FIBUA	fighting in built-up areas
GOC	General Officer Commanding

HiLOCA	high-level operations using cellular automata
IDP	internally displaced person
IO	information operations
IT	information technology
JSIMS	Joint Simulation System (US)
NATO	North Atlantic Treaty Organisation
NDM	naturalistic decision-making
NGO	non-government organisation
NTC	National Training Centre (US)
MC	Military Cross (United Kingdom)
MEL	mission events list
NOW	National Organization for Women (US)
O/C	Observer/Controller
OODA	observation–orientation–decision–action
OPFOR	opposing force
OR	other rank
PLO	Palestine Liberation Organisation
PTSD	post-traumatic stress disorder
PW	prisoner of war
RMA	Revolution in Military Affairs
RPA	Rwanda Patriotic Army
RPD	recognition-primed decision-making
RPF	Rwanda Patriotic Front
RTA	Restructuring the Army (Australia)
SA	situation awareness
SIMNET	simulation trials
SIT	stress inoculation technique
SNCO	senior non-commissioned officer
SNS	sympathetic nervous system
TES	tactical engagement simulation
TRACES	tactically relevant assessment of combat events
UAV	unmanned aerial vehicle
UK	United Kingdom
UN	United Nations
UNAMIR	United Nations Assistance Mission in Rwanda
UNIFIL	United Nations Interim Force in the Lebanon
UNPROFOR	United Nations Protection Force
USA	United States of America
USMC	United States Marine Corps
VC	Victoria Cross (United Kingdom)
VDU	video display unit
WEL	Women's Electoral Lobby (Australia)

Contributors

DR HELEN BRAITHWAITE

Helen Braithwaite received her PhD in psychology from Flinders University of South Australia in 1998, where she studied conflict resolution tactics of police officers on patrol. The research was published in a book, and she has published journal articles and book chapters on the subject.

Dr Braithwaite has qualifications in criminology, and has worked as an Intelligence Analyst with the National Crime Authority. She joined the Human Factors Discipline of Land Operations Division, DSTO, at Salisbury, in 1998. She has experience as the Staff Officer Science to 3 Brigade and is currently working on teamwork and collective training issues for the development of an Army Combat Training Centre.

PROFESSOR JOHN ENGLISH

John English was educated at Royal Roads Military Academy (1958–60) and the Royal Military College, Kingston, Canada (1960–62). He retired from the army as a lieutenant colonel after 35 years' service with Princess Patricia's Canadian Light Infantry, the Queen's Own Rifles of Canada and various Canadian and

British units in England, Germany, Denmark, Cyprus, Canada and Alaska. He also served as a NATO war plans officer, Chief of Tactics of the Combat Training Centre, and Directing Staff member of the Canadian Land Forces Command and Staff College. He holds an MA in history from Duke University (1964), an MA in war studies from Royal Military College, Kingston (1980), a PhD from Queen's University (1989), and is currently a professor of strategy with the US Naval War College.

Professor English is the author of *A Perspective on Infantry* (Praeger, 1981), republished as *On Infantry* (1984); *The Canadian Army and the Normandy Campaign: A Study of Failure in High Command* (1991); and *Marching Through Chaos: The Descent of Armies in Theory and Practice* (1996). He is also principal editor of *The Mechanized Battlefield: A Tactical Analysis* (1984), co-author of *On Infantry: Revised Edition* (1994), main Canadian contributor to the *D-Day Encyclopedia* (1994), and author of *Lament for an Army: The Decline of Canadian Military Professionalism* (1998). He has written numerous chapters for books, and his articles have appeared in *Military Affairs, Jane's Military Review*, the *Naval War College Review, Infantry*, the *Marine Corps Gazette*, the *Canadian Defence Quarterly*, and the proceedings of the Canadian Institute of Strategic Studies. He is adviser to the Praeger Series in War Studies and is researching another book, to be called *Omaha Beach*.

DR MICHAEL EVANS

Michael Evans is a Senior Research Fellow in the Land Warfare Studies Centre, Duntroon. He was formerly command historian at Land Headquarters, Sydney. He is a graduate in history and war studies of the universities of Rhodesia, London and Western Australia. Dr Evans has been a Beit Fellow in the Department of War Studies at King's College, University of London, and a Visiting Fellow at the University of York in England.

Dr Evans did national service in the Rhodesian Army and was later a regular officer in the Zimbabwe Army; with the rank of Major, he headed that army's War Studies Program and worked with the British Army in developing a staff college. He has published journal articles and papers in Australia, Britain, South Africa and the USA, and has recently completed a study of the development of Australian Army doctrine since 1947.

DR PADDY GRIFFITH

Paddy Griffith was born in Liverpool, England, and went to university at Corpus Christi College, Oxford. In 1968 he wrote an undergraduate thesis on French assault tactics of the Revolutionary and Napoleonic wars, and in 1976 completed his doctorate on 'Military Thought in the French Army 1815–51' (published in 1989). In 1973 he took up a lecturing post in War Studies at the Royal Military Academy, Sandhurst. In 1989 he left Sandhurst to become a freelance writer and publisher.

Dr Griffith's publications include *Forward Into Battle* (1981, 2nd edn 1990), which examined the reality of battlefield tactics from Waterloo to the present day; *Rally Once Again*, published in the USA as *Battle Tactics of the Civil War* (1989); and *Battle Tactics on the Western Front 1916–18* (1994). This last work sought to redress the balance of historical writings which have portrayed German tactics as innovative and imaginative, and castigated British tactics as conservative and failing to exploit modern weapons. Dr Griffith has also published on topics as diverse as *The Viking Art of War* and *The Art of War in Revolutionary France*.

LIEUTENANT COLONEL DAVE GROSSMAN, US ARMY RETIRED

Dave Grossman was an officer in the US Army for over 20 years, including long service as a ranger. He is a graduate of the British Army Staff College, Camberley, and has served on the teaching staff at the United States Military Academy, West Point. Most recently he was Professor of Military Science at Arkansas State University.

Lieutenant Colonel Grossman has acted as a consultant instructor on the US Army's Special Operations Command Psychological Operations Course; the US Marine Corps Amphibious Warfare School; the Bureau of Alcohol, Tobacco and Firearms; the US Veterans' Administration; the Texas Rangers; and the state police of California, Texas, Delaware, Minnesota and Nebraska. He has been a visiting lecturer at West Point, Camberley, Quantico and the Royal Canadian Military Institute. He was also an expert witness in the Oklahoma City bombing trials. His most recent book, *On Killing: The Psychological Cost of Learning to Kill in War and Society* (1996), is in its sixth printing in the USA, and has recently been translated into Japanese and Italian. He is also a contributor to the *Oxford Companion to American Military History*.

DR ELEANOR HANCOCK

Eleanor Hancock has been a lecturer in German history in the History Department of Monash University since 1988. She was previously a tutor at the University College, University of New South Wales, and a diplomat in the Department of Foreign Affairs.

Dr Hancock is the author of *The National Socialist Leadership and Total War, 1941–1945* (1992), and is currently researching a biography of the Nazi leader Ernst Roehm. She also researches and writes on feminism and traditional high politics, including foreign and defence policy.

DR DAVID HORNER

David Horner graduated from the Royal Military College, Duntroon, in 1969 and served as an infantry platoon commander in Vietnam in 1971. After various regimental and staff appointments he graduated from the Australian Army's Command and Staff College in 1983. From 1985 to 1988 he was a visiting fellow in the Department of History, University College, University of New South Wales, Australian Defence Force Academy. From 1988 until he retired from the Regular Army as a lieutenant colonel towards the end of 1990 he was a member of the Directing Staff of the Joint Service Staff College. As an Army Reserve officer he was employed on various special projects for Army Headquarters and the Department of Defence, and in January 1998 was promoted to colonel and appointed head of the Australian Army's Land Warfare Studies Centre.

Dr Horner has an MA in military history from the University of New South Wales and a PhD in military history and strategic studies from the ANU. In 1976 he was awarded a Churchill Fellowship to investigate the study of military history overseas, and in 1981 was awarded the ANU's most prestigious PhD research prize. He is a Senior Fellow in the Strategic and Defence Studies Centre at the ANU. He is the author or editor of 20 books on Australian military history, strategy and defence, including *Crisis of Command* (1978), *High Command* (1982), *The Commanders* (1984),

SAS: Phantoms of the Jungle (1989), *Duty First* (1990), *General Vasey's War* (1992), *The Gulf Commitment* (1992), *The Gunners: A History of Australian Artillery* (1995), *Inside the War Cabinet: Directing Australia's War Effort, 1939–1945* (1996), *Breaking the Codes: Australia's KGB Network 1944–1950* (1998) and *Blamey: The Commander-In-Chief* (1998). He is currently completing a biography of Sir Frederick Shedden, the Secretary of the Australian Department of Defence from 1937 to 1956.

Dr Horner is the editor of the Australian Army's military biography series and is the Visiting Fellow in military history at the Army's Command and Staff College. He has been the historical consultant for various television programs and has lectured widely on military history and strategic affairs.

DR JEREMY MANTON

Jeremy Manton migrated to Australia with his family and completed his secondary schooling in Sydney. He completed his PhD in experimental psychology and started working on sonar display systems at the Royal Australian Navy Research Laboratory in Sydney. In the early 1980s he was attached to the Admiralty Marine Technology Establishment, working on combat information system design for surface ships. On his return from the UK he took up a position in the Human Factors Group at the Aeronautical Research Laboratory, later heading the group. He was involved with research into aircrew avionics interactions issues, such as heat stress research in the Sea King helicopter, the Sea Hawk OT&E, the P–3C Orion Update and the F–111 ALR 2002 Radar Warning Receiver programs.

After a period as the Attaché Defence Science in Washington, DC, Dr Manton returned in 1996 and took up the role of leading the Human Systems Integration Group in the DSTO's Information Technology Division. He developed tasks in the area of asset visualisation tools for commanders and modelling operator activities in off-board decoy deployment for the RAN. In April 1998 he was appointed Research Leader—Human Factors, in Land Operations Division. He was the inaugural Leader of the DSTO Human Factors Hub and is National Leader for the Technical Cooperation Programme in the area of Human Resources and Performance. He is a member (consultant) of the Australian Army Psychology Corps and has served with 1 Commando Company, 144 Field Ambulance (British Territorial Army) and Melbourne University Regiment.

DR ALAN RYAN

Dr Alan Ryan is a Research Fellow in the Land Warfare Studies Centre, Duntroon. He was formerly Assistant Dean in the Colleges of Law and Arts at the University of Notre Dame Australia where he was senior lecturer in history, politics and law. He graduated in Law and Arts from The University of Melbourne. From 1987 to 1991 Dr Ryan was the Australian Pembroke Scholar at Cambridge University where he completed a doctoral thesis entitled *Indefeasible state sovereignty, the international community and attempts to abrogate war.*

Dr Ryan served in the Australian Army Reserve between 1981 and 1994, spending four years on attachment with the British Territorial Army. He has published in the areas of international affairs; political and military history; and historical method. He is currently completing a work on the deployment of the multinational coalition to East Timor.

ASSOCIATE PROFESSOR HUGH SMITH

Hugh Smith is a member of the School of Politics at the University College, Australian Defence Force Academy. He obtained his first degree from the London School of Economics and Political Science. He completed his PhD at the ANU and joined the Faculty of Military Studies at the Royal Military College, Duntroon, as a lecturer in 1971. He was founding director of the Australian Defence Studies Centre from 1987 to 1991 and is currently director of its Peacekeeping Program.

Associate Professor Smith teaches postgraduate courses on armed forces and society and on legal and moral problems of international violence; and undergraduate courses on 'War in International Politics' and on the ideas of Clausewitz. He has published widely on such topics as defence policy, reserve forces, conscientious objection, women in the military, officer education and other personnel issues and has edited books on the military profession, peacekeeping, and law and armed forces. Together with Professor Ian McAllister of the ANU, he has been conducting the Survey of the Military Profession in Australia since 1987.

PROFESSOR ROGER SPILLER

Roger J. Spiller grew up in Texas and served in the US Air Force 1962–65 as an air rescue medic. In 1969 he graduated from South West Texas State University, and he gained his PhD in US military history from Louisiana State University in 1977.

In 1979 Professor Spiller was a founding member of the US Army's Combat Studies Institute (CSI) and served as Director CSI in 1990–91 and 1993–94. He is currently George C. Marshall Professor of Military History at the US Army Command and General Staff College, Fort Leavenworth, including teaching in the School of Advanced Military Studies. From 1982 to 1985 he was Special Assistant to the Commander in Chief of US Readiness Command in Tampa, Florida, and from 1991 to 1995 he was Special Assistant (and personal historian) to the US Army Chief of Staff.

Professor Spiller is general editor of the three-volume *Dictionary of American Military Biography* and a consultant editor for *American Heritage Magazine*. His most recent article is 'In the Shadow of the Dragon: Doctrine in the US Army after Vietnam', in the *RUSI Journal*, January–February 1998. He is also currently writing a book on the individual experience of battle, entitled *In Wartime*.

COLONEL STEPHEN TETLOW

Stephen Tetlow was commissioned into the Royal Electrical and Mechanical Engineers (REME) and has served in a variety of all-arms appointments, mainly in Germany, Northern Ireland and the United Kingdom. More recently, he commanded the Combat Service Support (CSS) Battalion of the Allied Command Europe (ACE) Mobile Force (Land), specialising in arctic warfare and serving in Norway, Italy, Turkey and Greece.

Colonel Tetlow has been a member of the Directing Staff at the Royal Military College of Science, Shrivenham, and Assistant Secretary to the British Army's Number 2 Appointments Board. In December 1996 he assumed his current appointment at the Directorate of Land Warfare in the UK Ministry of Defence, where he is responsible for concepts and analysis in support of force development and future warfighting technologies. He has attended the British Higher Command and Staff Course and is a chartered engineer. He was awarded the MBE on the Northern Ireland Operational Awards List in 1991.

BRIGADIER JIM WALLACE

Jim Wallace was born in Sydney but spent his childhood in Brisbane, attending the Balmoral State High School. He graduated from the Royal Military College, Duntroon, in 1973 into the Royal Australian Infantry. His initial regimental service was as a subaltern in the 8th/9th Battalion, the Royal Australian Regiment, and subsequently with Special Forces, including over eight years in the Special Air Service Regiment (SASR). He was the commanding officer of the regiment from 1988 to 1990 and Commander of the Australian Special Forces from 1993 to 1995. In 1984 he was appointed a member of the Order of Australia for his services to the development of the SASR counterterrorist capability.

Brigadier Wallace is a graduate of the British Army Staff College, Camberley, and the Australian College of Defence and Strategic Studies. Staff and training appointments have included being Staff Officer to the Chief of Operations–Army, and as an instructor at the British Army Staff College, Camberley. From 1996 to 1998 he served as Commander 1st Brigade before assuming his current appointment as Director-General Land Development in Australian Defence Headquarters in January 1999.

COLONEL P.G. WARFE

Peter Warfe is a military physician specialising in tropical and preventive medicine. He was born in Melbourne and was commissioned as an Army undergraduate medical student in 1972. He graduated MB BS from Monash University in 1974 and in 1989 completed a Masters degree in Public Health and Tropical Medicine at the Uniform Services University, Bethesda, Maryland.

Colonel Warfe has served in Papua New Guinea, in northwestern Europe with NATO forces, and in the USA. He is a graduate of the Australian Army's Command and Staff College, Fort Queenscliff, and has held six command appointments during his 26 years of service. In 1995 he acted as the Force Medical Officer and commanded the Australian Contingent to the Second United Nations Assistance Mission in Rwanda (UNAMIR II). He was awarded the Conspicuous Service Cross for outstanding achievement as the contingent commander. Colonel Warfe is currently serving as the Director of Clinical Policy in Australian

Defence Headquarters. He is the author of numerous papers on military medicine issues, stress and peacekeeping.

DR CARLENE WILSON

Carlene Wilson is an organisational psychologist with over 30 publications in the areas of cognitive, organisational and social psychology. She completed her PhD at the University of Adelaide in 1984. After teaching assessment and vocational psychology at Deakin University for a number of years she joined the National Police Research Unit (now the Australasian Centre for Policing Research) in 1990, becoming Principal Researcher in 1991. Her work included examining ways of improving police performance both operationally and organisationally and culminated in co-authoring the book *Psychology and Policing* in 1996. Dr Wilson recently joined Land Operations Division of the Defence Science and Technology Organisation as a Senior Research Scientist, where she has researched aspects of workgroup functioning and command and control decision making.

Preface

IN MARCH 1999, *Land Warfare Doctrine 1: The Fundamentals of Land Warfare* enunciated the Australian Army's keystone doctrine into the 21st century. This publication expressed the Army's view that, although future warfare in the information age will be waged in a lethal battlespace with advanced technology, combat itself will retain its essential and age-old human features.

It was fitting, then, that three weeks after the publication of the Army's new keystone doctrine the Australian Army's 'think tank', the Land Warfare Studies Centre, should hold its inaugural international conference, 'The Human Face of War: Past, Present and Future'. The conference brought together an interdisciplinary group of eminent Australian, American, British and Canadian defence scholars and military practitioners from both academia and the armed forces. Their papers covered a wide range of topics, ranging from the psychology of combat and the experience of front-line soldiering to such issues as the role of women in combat, the impact of casualties in democratic societies, post-traumatic stress management, and the human effects of recent developments in digitisation. The conference was a remarkable success and was distinguished throughout by the high quality and diversity of its presentations.

The Australian Army has since deployed to East Timor in the largest commitment of troops since the Vietnam War. The

Army's experience in East Timor vindicates the emphasis we have placed on the importance of human factors in contemporary military operations. The central theme in this collection of conference proceedings is that a modern army's most precious asset remains the quality of its soldiers. It is a sentiment I endorse wholeheartedly. I recommend this informative and illuminating book to all those seeking to gain more knowledge of the human dimension of modern conflict.

F. J. HICKLING, AO, CSC
Lieutenant General
Chief of Army

Introduction

Roger J. Spiller

ON A CERTAIN day late in March 1999, the auditorium at Canberra's National Conference Centre was filled to capacity with delegates from the whole of the Australian defence establishment. There were high-ranking public servants, professional officers and non-coms of all ranks and services, defence analysts both official and freelance, scientists and physicians, and academics from several different tribes. Some foreigners, of whom I was one, basked happily in welcoming hospitality. My first impression was that I was in the company of serious minds on serious business, and as the next two days went on, that impression was wholly confirmed.

As I looked about the room I wondered why it was that I and my fellow outlanders had to go halfway round the world to find a professional conference that addressed the fundamental question of how war and lesser forms of conflict tested those who were charged with their conduct. I thought then, and have had no reason to unthink it since, that surely my own defence establishment should be as interested in this question as any other. But there was no official representative from any of my own armed forces. I was a guest of the Australian Army's Land Warfare Studies Centre, the research group that had organised the conference at the direction of the Chief of Army, Lieutenant General Frank Hickling.

The reasons for my own country's official absence, and the reasons for Australia's interest in the subject at hand, are worth considering. It can come as no surprise to students of modern defence affairs that my own armed forces and a good many others have committed themselves rather enthusiastically to a materialist, instrumental conception of war. This conception requires that, wherever and whenever possible, direct human action will be kept to an absolute minimum. I do not think that this conception of war arises from what Clausewitz might have called 'misplaced humanism'. I see no evidence that the leading armed forces have suddenly acquired a new regard for the sanctity of human life.

In my own country and in other leading democracies, techno-logical advances seem to promise that fine calculations can be made of what Dr Hugh Smith calls in one of the essays published here, the 'bearable cost' of any conflict; that is, an estimate of just how many friendly casualties any given military operation might be worth to the nation's interests. No doubt, many of those who advocate this technocentric view believe that it is the natural result of remarkable technological progress over the past several decades, and that this conception of modern conflict is simply the most reasonable, militarily expedient approach to the exercise of national power.

Instead, I think, the true origins of this technocentric view lie elsewhere, in a certain social outlook shared by many profes-sionals in my own armed forces and in those of other leading nations as well. It is an outlook that stands in harsh judgement on the human material of the modern societies whose interests they defend, and can be understood as a belief that the will of the modern citizen is too fragile to support any but the most benign military action. That this prejudice is not only ungenerous but also quite wrong seems not to have penetrated the official mind, at least in my own country. I believe that, in fact, the design of NATO's recent operations against Yugoslavia—which, coincidentally, began on the same day as the conference—was directly conditioned by just this prejudice.

Advocates of technocentric war need not express their pre-judice directly, however. A genuine, far-reaching military–technical revolution whose consummation lies well in the future is suffi-ciently advanced to imagine that near-bloodless war is within reach, and that if direct human action is not quite obsolete, it is certainly obsolescing. Merely pointing toward the advantages of what is called 'technological overmatch' seems to do the trick. One look at the balance between human and material investments in military budgets the world over is sufficient to make the point. The materialist conception of war seems to be winning out.

One might ask why it shouldn't. If an armed force has the means to accomplish by material what would otherwise be won at the cost of casualties, who but the bloodthirsty could argue against such an approach? The only reasonable answer, it seems to me, is that war has not yet organised itself to suit our preferences, and that while the near-bloodless war is an appealing vision, the realities of modern war tell us that it is still a fundamentally human act, that it is among humans that we still find both the causes and solutions for it. What John Keegan famously called 'the face of battle' is still the human face itself.

This simmering tension between the material and human conceptions of war, modern polemicists would have us believe, is of recent vintage, perhaps the offspring of cultural and techno-logical gyrations since the Vietnam War. But well before the birth of this century, voices could be heard wondering if the modern industrial citizen was equal to the demands of the modern indus-trial, ever more destructive, battlefield. On the eve of the Franco-Prussian War, Colonel Ardant du Picq held that a modern soldier had to withstand a greater terror in a shorter time than any of his forebears had to do. Twenty years later, the Polish banker and sometime military theorist, Jan de Bloch, produced a six-volume work that answered a resounding 'no', and concluded that war had therefore lost any utility it might once have had.

These doubts were not confined to narrow military circles. At about the same time Bloch was producing his massive study, the American novelist Stephen Crane produced by sheer imagination as succinct an expression of these concerns as one could find anywhere. 'Greek-like struggles were no more', Crane wrote in his classic, *The Red Badge of Courage*, 'Men were better, or more timid. Secular and religious education had effaced the throat-grappling instinct, or else firm finance held in check the passions.' One might have thought that the experience of the 20th century would have dispelled these suspicions, but one hundred million casualties later, they have not. Indeed, it is possible to see traces of this very question—are we still up to it?—between the lines in several essays published here.

Seen as a whole, this collection of essays, each presented at the conference and vigorously debated during open forum, says a good deal about the nature of the subject itself. Particularly in the last two decades, a body of literature has grown up that addresses the human face of warfare from any number of directions. Scholars from fields as disparate as cultural and social history, political science and international relations, sociology, psychology, general medicine and computer science have made their contributions, but this subject has remained resolutely non-theoretical, interdisciplinary. No one approach has attained sufficient breadth to fuse together these different branches of knowledge to create a comprehensive theory of war at the human level. Because of this, the open forums after each session were marked by a level of discussion as sophisticated and stimulating as any I have ever found at a professional conference.

These are discussions important for any armed force to have. Given Australia's unique history and military traditions, they were particularly important. The great, unifying traditions of the digger and the Anzac, of individualism and mateship, still spell themselves out in Australia's conception of war, its military doctrines and it operational practices. They explain why the Australian armed forces, as modern as they are, insist that of all their assets, the first are the Australians who fill their ranks.

For all that, the subjects under discussion did not turn merely on abstract questions, good only for pleasant speculation. For any army, no less the Australian Army itself, there stands behind all such meetings the prospect that one day, what is learned may be put directly to use. Though no one could have known at the time, seven months later the Australian armed forces would be conducting their largest overseas deployment since the end of the Vietnam War, and the subjects discussed at this conference would quickly take on a hard, practical edge. Many of the soldiers, airmen and sailors filling the halls with animated talk are no doubt on very active service as this is being written.

The mission now being conducted in East Timor is precisely the kind that defies the easy, technological solutions that are so blithely promoted in some quarters today. The conflict in East Timor has its origins in the human dimension, and that is where it will be played out. I hope that those who attended this conference and now find themselves directly engaged in this conflict found some profit in the conference whose proceedings are published here.

1 | Human factors in war: the psychology and physiology of close combat

Dave Grossman

THE PSYCHOLOGY AND physiology of close combat is a field that encompasses a wide variety of processes and negative impacts, all of which must be taken into consideration in any assessment of the immediate and long-term effects and costs of war. This chapter looks at the wide-spectrum effects of war on the individual in close combat, including psychiatric casualties suffered during combat, physiological arousal and fear, the physiology of close combat, the price of killing, and post-traumatic stress disorder (PTSD).[1]

INTRODUCTION: A LEGACY OF LIES

One obvious and tragic price of war is the toll of death and destruction. But there is an additional effect, a psychological cost borne by the *survivors* of combat, and a full understanding of this cost has been too long repressed by a legacy of self-deception and intentional misrepresentation. After peeling away this 'legacy of lies' which has perpetuated and glorified warfare there is no escaping the conclusion that combat, and the killing that lies at the heart of combat, is an extraordinarily traumatic and psychologically costly endeavour which profoundly affects all that participate in it.

This psychological and physiological effect of close combat is most readily observable and measurable at the individual level. At

5

the national level, a country at war can anticipate a small but statistically significant increase in the domestic murder rate, probably due to the glorification of violence and the resultant reduction in the level of 'repression' of natural aggressive instincts which Freud held to be essential to the existence of civilisation. At the group level, even the most elite unit is usually psychologically destroyed when 50–60 per cent casualties have been inflicted, and the integration of the individual into the group is so strong that this destruction often leads to depression and suicide. However, the nation (if not eliminated by the war) is generally resilient, and the group (if not destroyed) is inevitably disbanded. The individual that survives combat may well end up paying a profound psychological cost for a lifetime. The cumulative impact of these effects on hundreds of thousands of veterans has the potential to affect society at large profoundly.

PSYCHIATRIC CASUALTIES IN WAR

Richard Gabriel has noted: 'Nations customarily measure the "costs of war" in dollars, lost production, or the number of soldiers killed or wounded'. But 'rarely do military establishments attempt to measure the costs of war in terms of individual suffering. Psychiatric breakdown remains one of the most costly items of war when expressed in human terms'.[2] Indeed, for the combatants in every major war fought in this century, there has been a greater probability of becoming a psychiatric casualty than of being killed by enemy fire.

A psychiatric casualty is a combatant that is no longer able to participate in combat due to mental (as opposed to physical) debilitation. Psychiatric casualties seldom represent a permanent debilitation, and with proper care they can be rotated back into the line. (However, Israeli research has demonstrated that, after combat, psychiatric casualties are strongly predisposed towards the more long-term and more permanently debilitating manifestation of post-traumatic stress disorder.)

The actual psychiatric casualty can manifest itself in many ways, ranging from affective disorders to somatoform disorders, but the treatment for the many manifestations of combat stress involves simply removing the soldier from the combat environment. The problem is that the military does not want simply to return the psychiatric casualties to normal life—it wants to return them to combat. These casualties can be understandably reluctant to do so.

The evacuation syndrome is the paradox of combat psychiatry. A nation must care for its psychiatric casualties as they are of no value on the battlefield (indeed, their presence in combat can have a negative impact on the morale of other combatants), and they can be used again as valuable seasoned replacements once they have recovered from combat stress. If combatants begin to realise that 'insanity' or instability is a ticket to evacuation, the number of psychiatric casualties will increase dramatically.

Continued 'proximity' to the battlefield (through forward treatment, usually within enemy artillery range) and an 'expectancy' of rapid return ('immediacy') to combat are the principles developed to overcome the paradox of the evacuation syndrome. These principles of proximity, expectancy and immediacy have proven themselves quite effective since World War I. They permit the psychiatric casualty to get the rest that is the only current cure for his problem, while not giving a message to still healthy comrades that insanity is a ticket away from the madness of the battlefield.

Even with the careful application of the principles of proximity, expectancy and immediacy, the incidence of psychiatric casualties is still enormous. During World War II, 504 000 men were lost from America's combat forces due to psychiatric collapse—enough to man 50 divisions. The USA suffered this loss despite efforts to weed out those mentally and emotionally unfit for combat by classifying over 800 000 men 4-F (unfit for military service) due to psychiatric reasons. At one point in World War II, psychiatric casualties were being discharged from the US Army faster than new recruits were being drafted in.

Swank and Marchand's World War II study of US Army combatants on the beaches of Normandy found that, after 60 days of continuous combat, 98 per cent of the surviving soldiers had become psychiatric casualties. And the remaining 2 per cent were identified as 'aggressive psychopathic personalities'.[3] Thus it is not too far from the mark to observe that: there is something about continuous, inescapable combat which will drive 98 per cent of all men insane; the other 2 per cent were crazy before they got there. Figure 1.1 is a schematic representation of the effects of continuous combat.

It must be understood that the kind of continuous, protracted combat that produces such high psychiatric casualty rates is largely a product of 20th century warfare. The Battle of Waterloo lasted

Figure 1.1 Effects of continuous combat

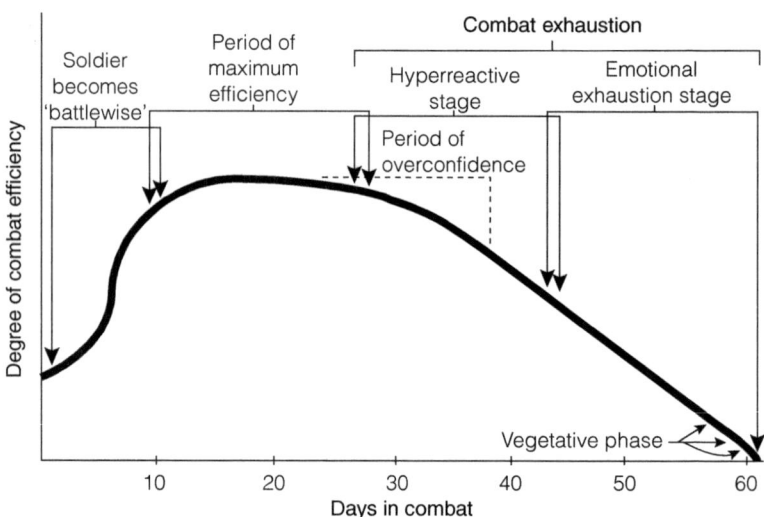

only a day. Gettysburg lasted only three days—and they took the nights off. It was only in World War I that armies began to experience continuous months of 24-hour combat, and it is in World War I that vast numbers of psychiatric casualties were first observed.

The democratic nations of this century have been better than most at admitting and dealing with their combat psychiatric casualties, and information from non-Western sources is extremely limited, but we now know that America's World War II experience is representative of a universal cost of modern, protracted warfare. Armies around the world have experienced similar mass psychiatric casualties, but many have simply driven these casualties into battle at bayonet point, shooting those who refused or were unable to continue. World War II Japanese units employed a unique set of powerful cultural and group processes to delay psychiatric breakdown, but they succeeded only in temporarily delaying the cost of combat, a cost which often manifested itself in mass suicide. Ultimately the toll of modern combat is truly fearful, and no nation or culture has been able to escape it.

PHYSIOLOGICAL AROUSAL AND FEAR

The soldier in combat endures many indignities. Among these can be endless months and years exposed to desert heat, sweltering jungle, torrential rains or frozen mountains and tundras. Usually the soldier lives amid swarming vermin. Very often there is lack of food, lack of sleep, and the constant uncertainty that eats away at the combatants' sense of control over their lives and their environment. Bad as they are, all of these stressors can be found in many cultural, geographic or social circumstances, and when the ingredient of war is removed, individuals exposed to these circumstances do not suffer mass psychiatric casualties.

To comprehend the intensity of the stress of combat fully we must keep these other stressors in mind while understanding the body's physiological response to combat, as manifested in the sympathetic nervous system's mobilisation of resources. Then we must understand the impact of the parasympathetic nervous system 'backlash' that occurs as a result of the demands placed on it.

The sympathetic nervous system (SNS) mobilises and directs the body's energy resources for action. It is the physiological equivalent of the front-line soldiers who actually do the fighting in a military unit. The parasympathetic nervous system (PNS) is responsible for the body's digestive and recuperative processes. It is the physiological equivalent of the body's cooks, mechanics and clerks, who sustain a military unit over an extended period of time.

Usually the body maintains itself in a state of homeostasis, which ensures that the sympathetic and parasympathetic nervous systems maintain a daily balance between demands on the body's resources. During extremely stressful circumstances the 'fight or flight' response kicks in and the SNS mobilises all available energy for survival. This is the physiological equivalent of throwing the cooks, mechanics and clerks into the battle. The reaction of the SNS is so intense that soldiers often suffer stress diarrhoea due to the redirecting of energies from non-essential parasympathetic processes, and it is not uncommon to lose control of urination and defecation as the body literally 'blows its ballast' and redirects all available energy in an attempt to provide the resources required for survival. This is reflected in World War II surveys, in which a quarter of combat veterans admitted that they urinated in their pants in combat, and about the same percentage admitted that they defecated in their pants in combat.

A combatant must pay a physiological price for such an enervating and intense process. The 'price' the body pays is an equally powerful 'backlash', when the neglected demands of the parasympathetic nervous system become ascendant. This parasympathetic backlash occurs as soon as the danger and the excitement are over, and takes the form of a powerful weariness and sleepiness on the part of the soldier. Napoleon stated that the moment of greatest danger was the instant immediately after victory, and in so saying he demonstrated an understanding of the way in which soldiers become physiologically and psychologically incapacitated by the parasympathetic backlash that occurs as soon as the momentum of the attack has halted and the soldier briefly believes himself to be safe. During this period of vulnerability a counterattack by fresh troops can have an effect out of all proportion to the number of troops attacking.

It is for this reason that the maintenance of a 'unblown' reserve has historically been essential in combat, with battles often revolving around which side can hold out and deploy its reserves last. Clausewitz understood the danger of reserve forces becoming prematurely enervated and exhausted (and provided insight into the root cause of the enervation) when he cautioned that the reserves should always be maintained out of sight of the battle.

In continuous combat the soldier rollercoasters through a seemingly endless series of surges of adrenalin and their subsequent backlashes, and the body's natural, useful and appropriate response to danger ultimately becomes counterproductive. Unable to flee, and unable to overcome the danger through a brief burst of fighting, posturing or submission, the bodies of modern soldiers in sustained combat exhaust their capacity to enervate and slide into a state of profound physical and emotional exhaustion.

Most observers of combat lump the impact of this physiological arousal process under the general heading of 'fear', but fear is really a cognitive or emotional label for non-specific physiological arousal in response to a threat. The impact of fear and its attendant physiological arousal is significant, but it must be understood that fear is just a symptom and not the disease; it is an effect but not the cause. To truly understand the psychological effects of combat, we must understand exactly what it is that causes this intense fear response in individuals, and it has become increasingly clear that there are two key, core stressors causing the psychological toll associated with combat. These stressors are: the trauma associated with being the victim of close-range, interpersonal aggression; and the trauma associated with being responsible for killing a fellow human being at close range.

THE TRAUMA OF CLOSE-RANGE, INTERPERSONAL AGGRESSION

During World War II the carnage and destruction caused by months of continuous German bombing in England, and years of Allied bombing in Germany, was systematically inflicted in order to create psychological casualties among civilian populations. Day and night, in an intentionally unpredictable pattern, for months and even years on end, relatives and friends were mutilated and killed, and homes were destroyed. These civilian populations suffered fear and horror of a magnitude that few humans will ever experience.

This unpredictable, uncontrollable reign of shock, horror and terror is exactly what psychiatrists and psychologists prior to World War II believed to be responsible for the vast numbers of psychiatric casualties suffered by soldiers in World War I. This mistaken belief laid the theoretical foundation for the German and Allied strategic bombing campaigns in World War II. Thus it came as a significant shock when the Rand Corporation's Strategic Bombing Study published in 1949 found that there was only a very slight increase in the psychological disorders in these populations as compared to peacetime rates, and that these occurred primarily among individuals already predisposed to psychiatric illness. These bombings, which were intended to break the will of the population, appear to have served primarily to harden the hearts and strengthen the determination to fight among those that endured them.

The impact of fear, physiological arousal, horror and physical deprivation in combat should never be underestimated, but it has become clear that other factors are responsible for psychiatric casualties among combatants. One of those factors is the impact of close-range, interpersonal, aggressive confrontation. Through roller-coasters, action and horror movies, drugs, rock climbing, white water rafting, scuba diving, parachuting, hunting, contact sports, and a hundred other means, modern society pursues fear. Fear in and of itself is seldom a cause of trauma in everyday peacetime existence, but facing close-range interpersonal aggression and hatred from fellow citizens is an experience of an entirely different magnitude.

The ultimate fear and horror in most modern lives is to be raped, tortured or beaten; to be physically degraded in front of loved ones or to have the sanctity of the home invaded by aggressive and hateful intruders. The Diagnostic and Statistical Manual of the American Psychiatric Association affirms this when it notes that post-traumatic stress disorder (PTSD) 'may be especially severe or longer lasting when the stressor is of human design'. PTSD resulting from natural disasters such as tornadoes, floods and hurricanes is comparatively rare and mild, but acute cases of PTSD will consistently result from torture or rape. Ultimately, like tornadoes, floods and hurricanes, bombs from 20 000 feet are simply not 'personal' and are significantly less traumatic—to both victim and aggressor.

Death or debilitation is statistically far more likely to occur by disease or accident than by malicious action, but statistics have nothing to do with fear. Statistically speaking, cigarette smoking is an extraordinarily dangerous activity which annually inflicts slow, hideous deaths on millions of individuals worldwide, but this fact does not dissuade millions of individuals from smoking, and around the globe few nations are motivated to pass laws to protect their citizens from this threat. But the presence of one serial rapist in a large city can change the behaviour of hundreds of thousands of individuals, and there is a broad tradition of laws designed to protect citizens from rape, assault and murder.

When snakes, heights or darkness cause an intense-fear reaction in an individual it is considered a phobia, a dysfunction, an abnormality. However, it is natural and normal to respond to an attacking, aggressive fellow human being with a phobic-scale response. This is a universal human phobia. More than anything else in life, it is intentional, overt, *human* hostility and aggression which assaults the self-image, sense of control and, ultimately, the mental and physical health of human beings.

The soldier in combat is inserted straight into the inescapable midst of this most psychologically traumatic of environments. Ultimately, if unable get some respite from the trauma of combat, and if not injured or killed, the combatant's only available escape is the psychological escape of becoming a psychiatric casualty and mentally fleeing the battlefield.

The physiology of close combat

An understanding of the stress of close combat begins with an understanding of the physiological response to close-range interpersonal aggression. The traditional view of combat stress is most often associated with combat fatigue and PTSD, which are actually manifestations that occur after, and as a result of, combat stress. The debilitating effects of combat stress have been recognised for centuries. Phenomena such as tunnel vision, auditory exclusion, the loss of fine and complex motor control, irrational behaviour and the inability to think clearly have all been observed as by-products of combat stress. Even though these phenomena have been observed and documented for hundreds of years, little research has been conducted to understand why performance deteriorates under combat stresses.

The key characteristic distinguishing combat stress is the activation of the SNS, which happens any time the brain perceives a threat to survival, resulting in an immediate discharge of stress hormones. This 'mass discharge' is designed to prepare the body for fight or flight. The response is characterised by rising arterial pressure and blood flow to large muscle mass (resulting in enhanced strength capabilities and gross motor skills, such as running from or charging into an opponent); vasoconstriction of blood vessels in the appendages (which serves to reduce bleeding from wounds); pupil dilation; cessation of digestive processes; and muscle tremors. Figure 1.2 is a schematic representation of the effects of hormonally induced heart rate increase resulting from SNS activation.

The activation of the SNS is automatic and virtually uncontrollable. It is a reflex triggered by the perception of a threat. Once initiated, the SNS will dominate all voluntary and involuntary systems until: the perceived threat has been eliminated or escaped; performance deteriorates; or the parasympathetic nervous system activates to re-establish homeostasis.

The degree of SNS activation centres around the level of perceived threat. For example, low-level SNS activation may result from the anticipation of combat. This is especially common with police officers or soldiers minutes before they make a tactical assault into a potential deadly force environment. Under these conditions combatants will generally experience increases in heart rates and respiration, muscle tremors, and a psychological sense of anxiety. In contrast, high-level SNS activation occurs when combatants are confronted with an unanticipated deadly force threat and the time to respond is minimal. Under these conditions the extreme effects of the SNS will cause catastrophic failure of the visual, cognitive and motor control systems.

Once activated, the SNS causes immediate physiological changes, of which the most noticeable and easily monitored is increased heart rate. SNS activation can drive the heart rate from

Figure 1.2 Effects of hormonal-induced heart rate increase

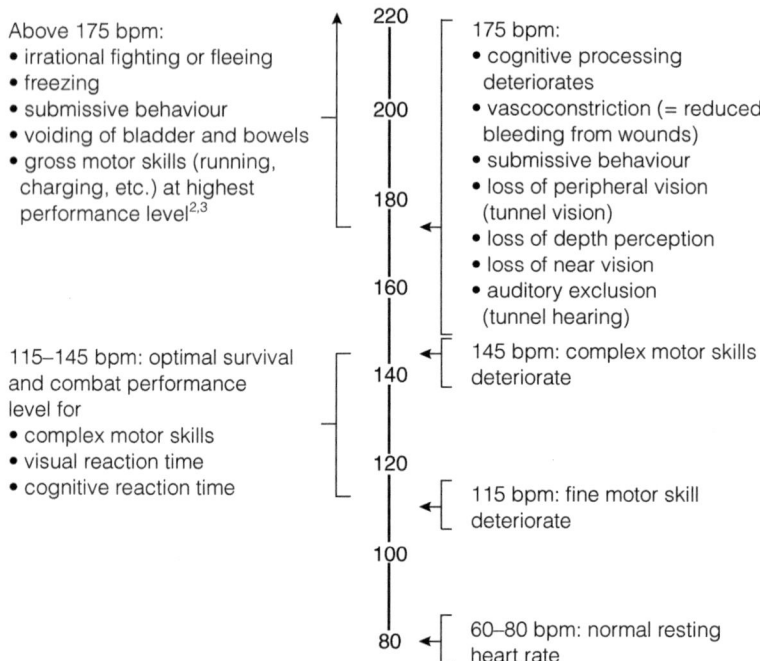

Heart rate (beats per minute)

Above 175 bpm:
- irrational fighting or fleeing
- freezing
- submissive behaviour
- voiding of bladder and bowels
- gross motor skills (running, charging, etc.) at highest performance level[2,3]

175 bpm:
- cognitive processing deteriorates
- vasoconstriction (= reduced bleeding from wounds)
- submissive behaviour
- loss of peripheral vision (tunnel vision)
- loss of depth perception
- loss of near vision
- auditory exclusion (tunnel hearing)

115–145 bpm: optimal survival and combat performance level for
- complex motor skills
- visual reaction time
- cognitive reaction time

145 bpm: complex motor skills deteriorate

115 bpm: fine motor skill deteriorate

60–80 bpm: normal resting heart rate

[1] These data are for hormonally induced heart rate increases resulting from sympathetic nervous system arousal. Exercise-induced increases will not have the same effect.

[2] Hormonally induced performance and strength increases can achieve 100% of potential max within 10 seconds, but drop to 55% after 30 seconds, 35% after 60 seconds, and 31% after 90 seconds. It takes a minimum of three minutes of rest to 'recharge' the system.

[3] Any extended period of relaxation after intense sympathetic nervous system arousal can result in a parasympathetic backlash, with significant drops in energy level, heart rate and blood pressure. This can manifest itself as normal shock symptoms (dizziness, nausea and/or vomiting, paleness, clammy skin) and/or profound exhaustion.

Source: Siddle & Grossman 1997.

an average of 70 to over 200 beats per minute in less than a second. As combat stress increases, heart rate and respiration will also increase until threat elimination or escape, catastrophic failure, or until the parasympathetic nervous system is triggered.

In 1950, S.L.A. Marshall's *The Soldier's Load and the Mobility of a Nation* was one of the first studies to identify how combat performance deteriorates when exposed to combat stress.[4] Marshall concluded that we must reject 'the superstition that under danger men can be expected to have more than their normal powers, and that they will outdo their best efforts simply because their lives are in danger'. Indeed, in many ways, the reality is just the opposite, and individuals under stress are far less capable of doing anything other than blindly running from or charging towards a threat. Humans have three primary survival systems: vision, cognitive processing and motor-skill performance. Under stress, all three break down.

Alexis Artwohl, a police psychologist in Portland, Oregon, has conducted post-combat interviews with police officers. In her research she has found that almost nine out of 10 of her subjects experienced diminished sound (in some cases so severe that a shotgun going off in front of an officer's face was not even heard, and did not cause a ringing in the ears afterward), while nearly two out of 10 experienced intensified sounds—usually occurring in night combat situations. Eight out of 10 experienced tunnel vision, and more than six out of 10 experienced 'slow motion time' and heightened visual clarity.[5] Research continues in this area, but it is increasingly obvious that there is a profound and severe set of physiological responses to combat, responses that have previously gone largely unsuspected (see Table 1.1).

Bruce K. Siddle's landmark research at PPCT Management Systems Inc. involved monitoring the heart rate responses of law enforcement officers in interpersonal conflict simulations using paintball-type simulation weapons. This research has consistently recorded heart rate increases to well over 200 beats per minute, with brief heart rates 'spikes' of up to 300 beats per minute.[6] These simulation pellets fired from real guns *hurt* when they hit, and thus in these simulations the combatants faced the universal human phobia—another human being who was trying to hurt them. Still, they knew that their life was not in danger. The difference between this and real combat is like the difference between a boxing match and a knife fight. The combatant in a true, life-and-death situation

Table 1.1 Perceptual distortions experienced in combat

88% Diminished sound (auditory exclusion)
82% Tunnel vision
78% Automatic pilot
63% Slow motion time
63% Heightened visual clarity
61% Memory loss for parts of the event
60% Memory loss for some of your actions
50% Dissociation (detachment)
36% Intrusive distracting thoughts
19% Memory distortions
17% Intensified sounds
17% Fast motion time
11% Temporary paralysis

Source: Dr Alexis Artwohl & Loren Christian, *Deadly Force Encounters*, Paladin Press, Boulder, CO, 1997. Based on post-combat surveys of 72 officers.

will probably experience a physiological reaction even greater than that of Siddle's subjects.

The fundamental truth of modern combat is that the stress of facing close-range interpersonal aggression is so great that, if enduring this for months on end without any other means of respite or escape, the combatant will inevitably become a psychiatric casualty. Even greater than the resistance to being the *victim* of close-range aggression is the combatant's powerful aversion to *inflicting* aggression on fellow human beings. At the heart of this dread is the average, healthy person's resistance to killing one's own kind.

A RESISTANCE TO KILLING

There is a notable reduction in the kind of psychiatric casualties usually identified with long-term exposure to combat among medical personnel, chaplains, officers and soldiers on reconnaissance patrols behind enemy lines. The key factor in each of these situations is that, although they are in the front lines and the enemy may attempt to kill them, they have no direct responsibility to personally participate in close-range killing activities. Even when there is equal or even greater danger of dying, combat is much less stressful if you do not have to kill.

The existence of a resistance to killing lies at the heart of this dichotomy between killers and non-killers. This is an additional, final stressor that the combatant must face. To understand the nature of this resistance to killing we must first recognise that most participants in close combat are literally 'frightened out of their wits'. Once the bullets start flying, the effects of vasoconstriction are such that blood flow to the forebrain begins to shut down: combatants stop thinking with the forebrain, which is the part of the brain that makes us human, and start thinking with the midbrain, or mammalian brain, the primitive part of the brain which is generally indistinguishable from that of any other mammal.

This process of the midbrain 'hijacking' of the forebrain is at the core of most severe combat effects. This process is to the combatant what the fundamentals of combustion and back draft are to the firefighter. In conflict situations this primitive, midbrain processing can be observed in the existence of a powerful resistance to killing one's own kind. During territorial and mating battles, animals with antlers and horns slam together in a relatively harmless head-to-head fashion, rattlesnakes wrestle each other, and piranha fight their own kind with flicks of the tail, but against any other species these creatures unleash their horns, fangs and teeth without restraint. This is an essential survival mechanism that prevents a species from destroying itself during territorial and mating rituals.

One major modern revelation in the field of military psychology is the observation that this resistance to killing one's own species is also a key factor in human combat. Brigadier General S.L.A. Marshall first observed this during his work as official historian in the Pacific and European theatres of operation in World War II. Based on his post-combat interviews, Marshall concluded in his landmark book, *Men Against Fire*, that only 15–20 per cent of the individual riflemen in World War II fired their weapons at an exposed enemy soldier. Soldiers using crew-served weapons such as machine guns almost always fired. Soldiers using key weapons, such as flame-throwers, usually fired. Firing would increase greatly if a nearby leader demanded that the soldier fire. When left to their own devices, however, the great majority of individual combatants throughout history appear to have been unable or unwilling to kill.[7]

Marshall's findings have been somewhat controversial. Faced with scholarly concern about a researcher's methodology and conclusions, the scientific method involves replicating the research. In Marshall's case, every available parallel scholarly study validates his basic findings of a powerful resistance in human beings to the close-range killing of their own species. Marshall's fundamental conclusion that man is not, by nature, a killer is confirmed by numerous studies. Those include Ardant du Picq's surveys of French officers in the 1860s and his observations on ancient battles, Keegan's and Holmes' numerous accounts of ineffectual firing throughout history, Richard Holmes' assessment of Argentine and British firing rates in the Falklands War, Paddy Griffith's data on the extraordinarily low killing rate among Napoleonic and American Civil War infantry regiments, the British Army's laser re-enactments of historical battles, the FBI's studies of non-firing rates among law enforcement officers in the 1950s and the 1960s, and countless other individual and anecdotal observations.[8]

The exception to this resistance can be observed in sociopaths, who, by definition, feel no empathy or remorse for their fellow human beings. Pit bull dogs have been selectively bred for sociopathy, bred for the absence of resistance to killing their own kind in order to ensure that they will perform the unnatural act of killing another dog in battle. Similarly, human sociopaths represent Swank and Marchand's 2 per cent that did not become psychiatric casualties after months of continuous combat, as they were not disturbed by the requirement to kill. However, sociopaths would be a flawed tool, impossible to control in peacetime, and social dynamics make it difficult for humans to breed themselves for such a trait. Humans *are* adept at finding mechanical means to overcome natural limitations. Born without the ability to fly, for example, we found mechanisms to overcome this limitation and enable flight. Humans also appear to have been born without the ability to kill fellow humans, and so, throughout history, we have devoted great effort to finding a way to overcome this resistance. From a psychological perspective, the history of warfare can be viewed as a series of successively more effective tactical and mechanical mechanisms to enable or force combatants to overcome their resistance to killing.

Overcoming the resistance to killing

By 1947 the US Army had accepted Marshall's conclusions, and the Human Resources Research Office of the US Army subsequently pioneered a revolution in combat training which eventually replaced firing at bullseye targets with deeply ingrained 'conditioning' using realistic, man-shaped pop-up targets that fell when hit. Psychologists know that this kind of powerful 'operant conditioning' is the only technique that will reliably influence the

Figure 1.3 Killing enabling factors

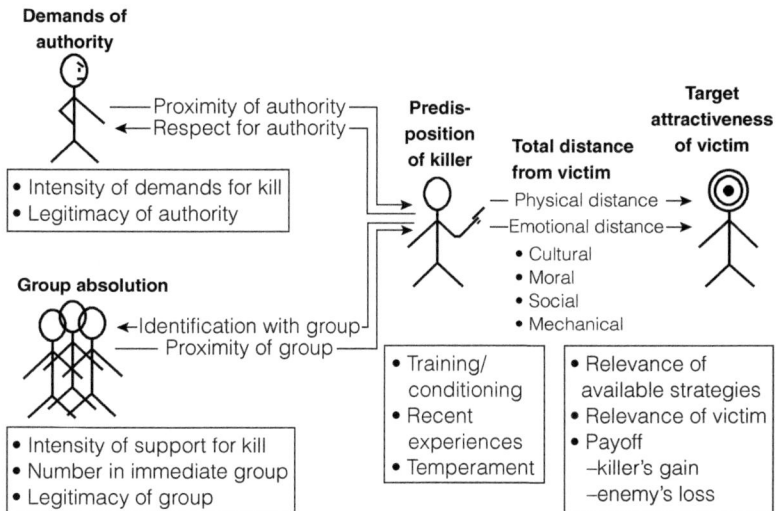

primitive, midbrain processing of a frightened human being, just as fire drills condition terrified school children to respond properly during a fire, and repetitious, 'stimulus–response' conditioning in flight simulators enables frightened pilots to respond reflexively to emergency situations.

Throughout history the ingredients of groups, leadership and distance have been manipulated to enable and force combatants to kill, but the introduction of conditioning in modern training was a true revolution. The application and perfection of these basic conditioning techniques increased the rate of fire from near 20 per cent in World War II to approximately 55 per cent in Korea and around 95 per cent in Vietnam. Similar high rates of fire resulting from modern conditioning techniques can be seen in FBI data on law enforcement firing rates since the nationwide introduction of modern conditioning techniques in the late 1960s. Figure 1.3 is a schematic representation of the interaction between the killing enabling factors which have been manipulated throughout history, including the key modern ingredient of conditioning.

One of the most dramatic examples of the value and power of this modern psychological revolution in training can be seen in Richard Holmes' observations of the 1982 Falklands War. The superbly trained (i.e. 'conditioned') British forces were without air or artillery superiority and were consistently outnumbered three to one while attacking the poorly trained but well-equipped and carefully dug-in Argentine defenders. Superior British firing rates (which Holmes estimates to be well over 90 per cent), resulting from modern training techniques, has been credited as a key factor in the series of British victories in that brief but bloody war. Any future army that attempts to go into battle without similar psychological preparation is likely to meet a fate similar to that of the Argentines.

The price of overcoming the resistance to killing

The extraordinarily high firing rate resulting from modern conditioning processes was a key factor in the USA's ability to claim that their ground forces never lost a major engagement in Vietnam. However, conditioning that overrides such a powerful, innate resistance carries with it enormous potential for psychological backlash. Every warrior society has a 'purification ritual' to help returning warriors deal with their 'blood guilt' and to reassure them that what they did in combat was 'good'. In primitive tribes this generally involves ritual bathing, ritual separation (which serves as a cooling-off and 'group therapy' session), and a ceremony embracing the veteran back into the tribe. Modern Western rituals traditionally involve long periods while marching or sailing home, parades, monuments, and the unconditional acceptance of society and family.

Table 1.2 outlines some of the key factors in the killing experience rationalisation and acceptance processes, using US troops in Vietnam as a case study of an extreme circumstance in which the purification rituals broke down. For example, combatants do not do what they do in combat for medals, they are motivated largely by a concern for their comrades; but after the battle medals serve as a kind of 'Get Out of Jail Free Card'— a powerful talisman that proclaims to them and to others that what the combatant did was honourable and acceptable. Although medals were issued in Vietnam, the social environment was such that veterans could not wear the medals or their uniforms in public. Similarly, the young combatant needs the presence of mature, older comrades from whom to seek guidance and support, but in the latter years of the war in Vietnam the average age of the combatant was 19, as opposed to 26 in World War II. Other key factors unique to the US experience in Vietnam include the absence of

Table 1.2 **Killing experience rationalisation and acceptance processes**

Process	Past wars	Vietnam
Praise from peers and superiors (medals, citations)	Yes	Yes (not worn)
The presence of mature, older comrades	Yes	No (reduced)
Circumstances limiting civilian kills/atrocities	Yes	No (reduced)
Rear lines and safe areas	Yes	No
Presence of close, trusted friends throughout the war	Yes	No
Cooling-off period with comrades while returning home	Yes	No
Knowledge of victory, gain, and accomplishments	Yes	No
Parades and monuments	Yes	No (delayed)
Reunions and continued communications with comrades after the war	Yes	No
Acceptance and praise from friends, family and society	Yes	No (mixed)
Support to veteran from religious and political systems	Yes	No (mixed)

any truly safe, secure area in-country; the individual replacement system, which hampered bonding and ensured that soldiers often arrived and left as strangers; and the use of aircraft to immediately return veterans to America, without the usual cooling-off, group therapy period experienced for thousands of years as veterans sailed or marched home.

For US Vietnam veterans the purification ritual was largely denied, and a large number of studies have demonstrated that one of the most significant causal factors in PTSD is the lack of support structure after the traumatic event, which in this case occurred when the returning veteran was attacked and condemned in an unprecedented manner. The traditional horrors of combat were magnified by modern conditioning techniques, and this combined with the nature of the war and an unprecedented degree of societal condemnation to create a circumstance which resulted in between 0.5 and 1.5 million cases (the results of studies vary greatly) of PTSD among the 3.5 million US veterans of South-East Asia. This mass incidence of psychiatric disorders among Vietnam veterans resulted in the 'discovery' of PTSD, a condition that has always occurred as a result of warfare.

Armies around the world have integrated these lessons from Vietnam; in Britain's Falklands War, Israel's 1982 Lebanon incursion, and the USA's Gulf War, the lessons of Vietnam and the need for a purification ritual have been closely and carefully considered and applied. In the former USSR's Afghanistan War this need was again ignored, and the resultant social turmoil was one of the factors that eventually led to the collapse of that nation. Indeed, the Weinberger Doctrine (later referred to as the Powell Doctrine), which holds that the USA will not engage in a war without strong societal support, is a reflection of the tragic lessons learnt from the psychological effects of combat in Vietnam.

PTSD is a psychological disorder resulting from a traumatic event. It manifests itself in persistent re-experiencing of the traumatic event, numbing of emotional responsiveness, and persistent symptoms of increased arousal, resulting in clinically significant distress or impairment in social and occupational functioning. There is often a long delay between the traumatic event and the manifestation of PTSD. Among Vietnam veterans in the USA, PTSD has been strongly linked to greatly increased divorce rates, higher incidence of alcohol and drug abuse, and higher suicide rates. Indeed, some studies indicate that, as of 1996, three times more Vietnam veterans had died from suicide since the war than from enemy action during the war, and this number is growing every year.

PTSD seldom results in violent criminal acts, and US Bureau of Justice statistics research indicates that veterans, including Vietnam veterans, are statistically less likely to be incarcerated than non-veterans of the same age. The key safeguard in this process appears to be the deeply ingrained discipline that the soldier internalises with military training. With the advent of interactive 'point-and-shoot' arcade and video games there is significant concern that society is aping military conditioning, but without the vital safeguard of discipline.

There is strong evidence to indicate that the indiscriminate civilian application of combat conditioning techniques as entertainment (specifically graphic visual displays of violence in television, movies and video games) may be a key factor in worldwide, skyrocketing violent crime rates, including a seven-fold increase in per-capita aggravated assaults in America since 1956, a five-fold increase in assault in Canada since 1964; and (according to Interpol data) the per-capita 'serious assault rate' between 1978 and 1993 has increased approximately five-fold in Norway and Greece, four-fold in Australia and New Zealand, three-fold in Sweden, and approximately doubled in a half-dozen other European nations (see

Table 1.3 International violent crime rates

	'Serious assault'			'Murder'		
	1977	1993	Increase	1977	1993	Increase
Australia[1]	21.9	81.3	+3.7x	2.8	4.5	+1.6x
Belgium	65.9	125.0	+1.9x	2.2	3.1	+1.4x
Canada[2]	447.0	916.0	+2.0x	3.0	2.0	—
Denmark	78.7	179.0	+2.3x	2.5	4.8	+1.9x
England/Wales[1]	163.0	362.0	+2.2x	1.4	2.5	+1.8x
France	59.8	99.0	+1.7x	3.4	4.9	+1.4x
Greece	14.4	68.4	+4.8x	1.2	2.5	+2.1x
Hungary[3]	45.1	76.9	+1.7x	3.5	4.5	+1.3x
Netherlands[4]	101.0	196.0	+1.9x	8.3	27.4	+3.3x
New Zealand[1]	83.4	313.0	+3.8x	1.8	4.0	+2.2x
Norway	12.8	62.0	+4.8x	0.7	2.5	+3.6x
Scotland[5]	53.0	123.0	+2.3x	8.4	11.4	+1.4x
Sweden	17.3	51.1	+3.0x	4.8	8.8	+1.8x
USA	241.0	440.0	+1.8x	8.8	9.5	+1.1x

1 Data are only through the following dates, when the indicated nations stopped reporting to Interpol: Australia 1988; England/Wales 1991; New Zealand 1992.

2 Canada does not report crime data to Interpol; Canadian data are from the Canadian Center for Justice.

3 Data begin in 1980, when Hungary started reporting to Interpol.

4 Netherlands did not begin reporting serious assault data to Interpol until 1981, but murder data begin in 1977.

5 Scotland's serious assault data begin in 1977, but murder data begin in 1985 (when they apparently started reporting murder under a broader definition) and both murder and serious assault data run only through 1991, when Scotland stopped reporting to Interpol.

Sources: All data represent incidents per 100 000 population, as reported by each nation to Interpol and recorded in Interpol International Crime Statistics, vols. 1977 to 1994 (except for Canadian data, as stated above in note 2). Different nations use different criteria to define 'murder' and 'serious assault', therefore the ability to use these data to compare between nations is limited, but comparisons of increases within each nation across time are valid. (This information was previously reported in a different format in *On Killing*, 1996, Dave Grossman.)

Table 1.3). Thus, the psychological effects of combat can increasingly be observed on the streets of nations around the world.

CONCLUSION: A CULTURAL CONSPIRACY

It is often said that 'All's fair in love and war'; this expression provides a valuable insight into the human psyche, as these taboo fields of sexuality and aggression represent the two realms in which most individuals will consistently deceive both themselves and others. Our psychological and societal inability to confront the truth about the effects of combat is the foundation for the cultural conspiracy of repression, deception and denial that has helped to perpetuate and propagate a highly unrealistic and potentially destructive image of the reality of war throughout recorded history.

In the field of developmental psychology, a mature adult is sometimes defined as someone that has attained a degree of insight and self-control in the two areas of sexuality and aggression. This is also a useful definition of maturity in civilisations. Thus two important and reassuring trends in recent years have been the development of the science of human sexuality, which has been termed 'sexology', and a parallel development of the science of human aggression, which I have termed 'killology'. There is clear consensus that continued research in this previously taboo realm of human aggression is vital to the future development, and perhaps to the very existence, of our civilisation.

2 | Incorporating human factors in simulation: a British Army view

Stephen Tetlow

IN 1945 BRIGADIER Nigel Balchin, a director to the British Army Council, commented that:

> The moral may be to the physical as three is to one, but the fact remains that at a time when literally hundreds of scientists were engaged in studying fragmentation and muzzle velocities, there was not a single man engaged full time in the study of those morale effects which are all that 95% of shells bombs and bullets produce.

At that time, as Brigadier Balchin suggested, only a small proportion of defence research examined the effect of human factors on combat performance. Little has changed. This chapter provides a general introduction to the steps the British Army has taken to rectify this by incorporating human factors in simulation to analyse the course of combat engagements.[1] The knowledge acquired from these simulations is employed in support of the British Army's force development process. The specific human variables covered are: fear; surprise; shock; participation in combat; and nationality factors. This chapter is intended to give a broad overview of the research; it is not intended to provide specific details of this research in great depth.

The father of military operational analysis, F.W. Lanchester, was an engineer who applied the strengths of his research discipline to the study of war. Lanchester's square law is a mathematical representation of concentration of force, which greatly enhanced the ability of the research community to support the armed services; it is used in operational analysis to support the acquisition of military systems to this day. Operational analysis supports a number of key elements of forward planning, most notably the design of the British Army in the future; the Army's assessments of defence capability, its campaign planning and its equipment acquisitions programs. However, operational analysis has retained a close allegiance to the physical sciences and, with a few notable exceptions, has largely avoided the less precise human sciences. Laudable efforts have been made over the years to improve the precision of operational analysis, but in many cases it has been what could be described as 'precisely wrong', simply because it has been unable to incorporate an understanding of the human aspects of combat.

WHY INCORPORATE HUMAN FACTORS IN SIMULATION?

The principal reason for wanting to include human factors in simulations is that we know they play a significant role in determining the outcome of combat. We have a genuine belief that simulations that neglect human factors are likely to have limited accuracy, and that incorporating the human aspects of conflict in our models is the right thing to do.

We consider that there are three basic components of fighting power, as shown in Figure 2.1. First, the moral component, which is essentially to do with the person—the ability to get people to fight. Second, the physical component, the main elements of which are equipment, manpower, logistics, training and readiness—the means to fight. Third, the conceptual component—the thought processes behind the ability to fight, such as the principles of war concepts and doctrine. The environment in which fighting power is applied influences all three, but it is the mixture and synergy of these components that determines fighting power. However, it is the physical component that is the most amenable to the application of operational analysis and to which most simulation has traditionally been applied. As Figure 2.1 attempts to illustrate, the moral and conceptual components have consequently largely been ignored.

Operational analysis may be logical and consistent, and it provides the necessary audit trail demanded of our programmers, but it is predominantly equipment-oriented. Without doubt, the

Figure 2.1 Military effectiveness

The three interrelated components of fighting power

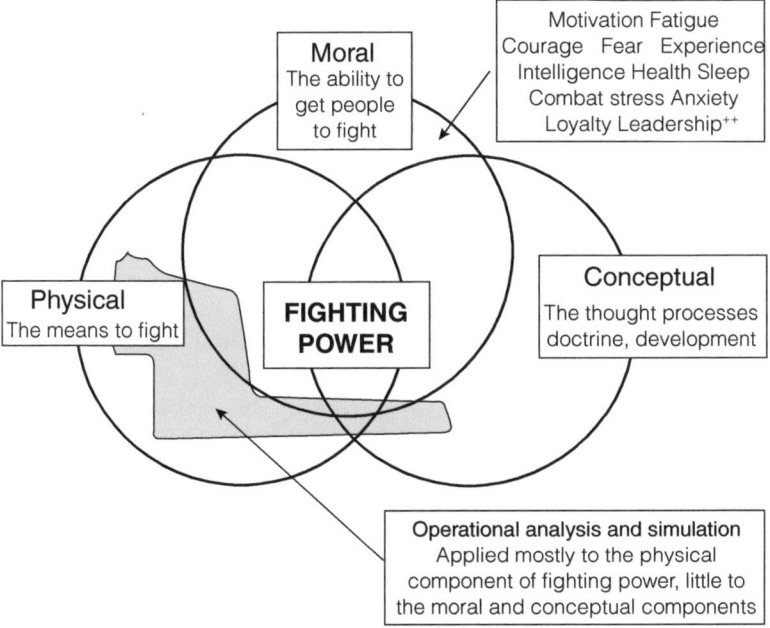

lack of representation of many human factors in simulations is due to the intrinsic complexity and difficulty of handling these factors analytically. Progress is not assisted by the fact that the research communities (i.e. the physical sciences, operational analysis, human sciences and historical research) are essentially separate disciplines.

We consider the incorporation of human factors in simulation and the bringing together of the underlying disciplines to be important for the foregoing reasons. Simulation informs so much of our force development and underpins a great deal of the analytical studies that are now a mandatory part of our weapon systems acquisition process. We ignore the human factors at our peril.

HOW WE INCORPORATE HUMAN FACTORS IN SIMULATION

The incorporation of human factors requires a multidisciplinary approached pitched at a number of different levels. First, we sponsor a Historical Co-ordination Group, which is intended to benefit all the research and development communities. The group has a strong human factors element and includes representatives of the operational analysis, historical analysis, systems assessment, human sciences and historical research communities, as well as analysts from some of our deployable headquarters. Regular meetings of the group assist in keeping the communities aware of activities in other areas. The forum has contributed considerably to the co-ordination and cross-fertilisation of Defence-wide research activity.

Second, we are influencing the efforts of the human sciences community to produce models or submodels that can be linked to operational analysis. So far, the nature of this work is speculative research to represent human factors in combat models (described later). Its purpose is to explore the issues involved in representing human factors in simulations supporting operational analysis and to expose these to a wider audience.

The third technique is the utilisation of the unique area of historical analysis, which studies a wide range of historical cases by scientific methods. The key drivers of this work are to do with human factors, involving the interaction between people and technology. (This is described in more detail later.) However, the results are used by the Army for decision support tools on operations, operational campaign planning, doctrine development, concepts and force development work, and are increasingly used for model development and model validation by the Defence analytical community.

It is worth spelling out the key elements of our historical analysis methodology. First, it is based on quantified primary source data—from both sides of conflict where possible. It is important to look at both sides because a lot of information that we receive is anecdotal, and the view from 'the other side of the hill' goes some way to restoring balance. Second, the analysis uses established statistical techniques to meet the requirements of the Ministry of Defence validators, who will ultimately have to accept it. Third, it focuses on understanding the underlying mechanisms—we try to get down deep into the human factors at the root of what we are trying to do. Fourth, as we are aware of the need to be pragmatic about what we are trying to do, data are quantified as far as is sensible. Fifth, there are two fundamental areas to ensure the credibility of the operational analysis model in validation: the pace of battle; and more importantly the actual level of performance in real conflict and the effect of human and organisational factors.

Finally, in view of the extremely important impact of forth-coming digitisation, we are attempting to improve the modelling of command, control, communications and information (C3I) in high-level operational analysis. The principal aim here is to allow command to be represented in fast-running, high-level aggregated force models focused on determining future force structures and operational combat capabilities. Another reason for this work is due to the failure of attempts to incorporate these factors in most of the current rule-based models, which has resulted in large, unfathomable and unwieldy rule bases. Our alternative approach is to create a set of relatively simple software-based 'command agents', which can take independent actions based on their views of goals and enemy forces. These have been named cellular automata, and the experimental platform for these is termed HiLOCA (high-level operations using cellular automata).

HUMAN FACTOR VARIABLES THAT NEED TO BE CONSIDERED

As can be seen, we are tackling the complex business of incorpo-rating human factors in simulation in several ways. The prime example of what we are doing relates to fear. We asked our human sciences community to investigate which factors might be the most appropriate to link to operational analysis techniques from their perspective. The first factor to emerge from this investigation was fear, and from the results of this work we developed a model of fear in combat. This was undertaken as a piece of speculative research in order to obtain some idea of the feasibility and utility of representing explicitly a complex human factor in an operational analysis model. Our fear model was developed as a stand-alone submodel with a clearly defined interface to our existing, systems dynamics-based, land combat model. There are two outputs from the fear model: first, a modifier to combat effectiveness; second, a modifier to a unit's defeat level and surrender levels. The latter are the nominal residual strengths of a unit below which it is deemed that unit would be defeated. The model embraces a number of factors that can be grouped under the following head-ings: enemy fire; casualties; level of cover; morale and cohesion; and isolation.

Figure 2.2 Top-level influence diagram in the fear model

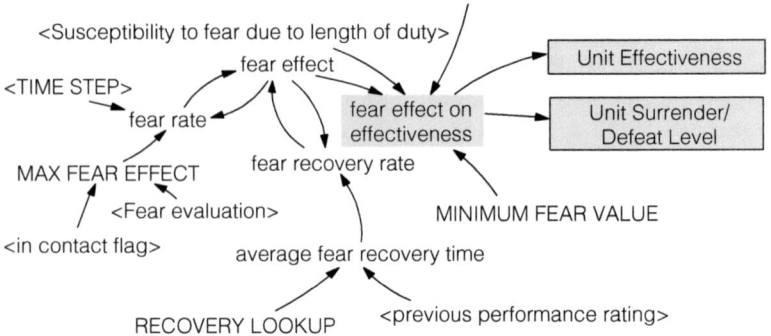

Our existing combat model is systems dynamics–based, which was a very important consideration in choosing the programming environment for this particular model. Figure 2.2 shows the top-level influence diagram in the model. Apart from the boxed descriptions (dark grey) in the diagram, all words and phrases are variables used within the model. The arrows between them simply demonstrate a direct influence. The 'recovery lookup' variable is a lookup graph. The influences in upper case are constants. Not shown here, but behind each variable, is an equation that defines the exact nature of the influence. The variables that appear with '< >' surrounding them (i.e. <fear due to enemy fire>) denote that these variables are calculated elsewhere in the model—there are other more detailed influence diagrams for each of these particular variables.

The level of fear being experienced is calculated at 'fear effect'. This increases at a rate determined at 'fear rate', which in turn is influenced by the drivers of fear which are calculated at <Fear evaluation>. Fear is assumed to decrease at a rate determined by 'fear recovery rate'. The extent to which the level of fear being experienced can influence effectiveness is determined by 'susceptibility to fear' within defined maximum and minimum boundaries determined by 'MAX FEAR EFFECT' and 'MINIMUM FEAR VALUE'. Plotted on a graph, different types of output can be obtained from the fear model when being run as a stand-alone model. Examples of such output might include representations of fear being experienced over time; the causes of fear (such as fear due to casualties, enemy fire, isolation, and levels of morale and

Table 2.1 Attack success in land campaigns

Dependence on surprise, air superiority and aggressive ground recce (excluding force ratios)

	Probability of attack breakthrough	Probability of attack campaign success
If none of the three applies	<5%	10%
If all three apply	>95%	90%

cohesion); or the relationship between a unit's strength and its defeat level.

The second key area of research concerns the effects of surprise and its significance to combat outcome. This has been studied extensively by the Historical Analysis Branch of the UK's Centre for Defence Analysis (CDA). From analysis of 160 land campaigns of the 20th century, this research has demonstrated that surprise ranks as one of the top three factors for success by an attacker (the other factors being air superiority and aggressive ground reconnaissance). If none of these conditions existed, neither breakthrough nor eventual campaign success was likely—indeed, there was a less than 10 per cent probability of either actually happening. Conversely, if all three conditions applied the success rate was likely to exceed 90 per cent. These results are displayed in Table 2.1.

A small but specific example of the extensive research carried out by CDA is illustrated in Figure 2.3. It demonstrates the effect of surprise on attacker casualties as a function of force ratio for infantry combat in open terrain, and quantifies it for modelling purposes. For example, taking a value of 1.0 with a force ratio of 3:1 in the attacker's favour, this graph shows no decrease in casualties achieved by surprise if the attacker has a force ratio in excess of 3:1. If the attacker force ratio exceeds that, the chances of achieving surprise decrease markedly. Additionally the data demonstrate that the benefit to the attacker of surprise depends on force ratio, for example, at 1:1 force ratio the benefit is to suffer 58 per cent of the casualties that would have been incurred without surprise.

Further work by CDA has quantified the combat significance of 'up-front' leadership by statistical analysis of heroic acts (i.e. acts that resulted in a gallantry award or citation). For instance, we have been able to measure a strong statistical correlation between defensive performance and incidence of heroic acts. This analysis has assisted our force development process by establishing and

Figure 2.3 Effect of surprise on attacker casualties with force ratio

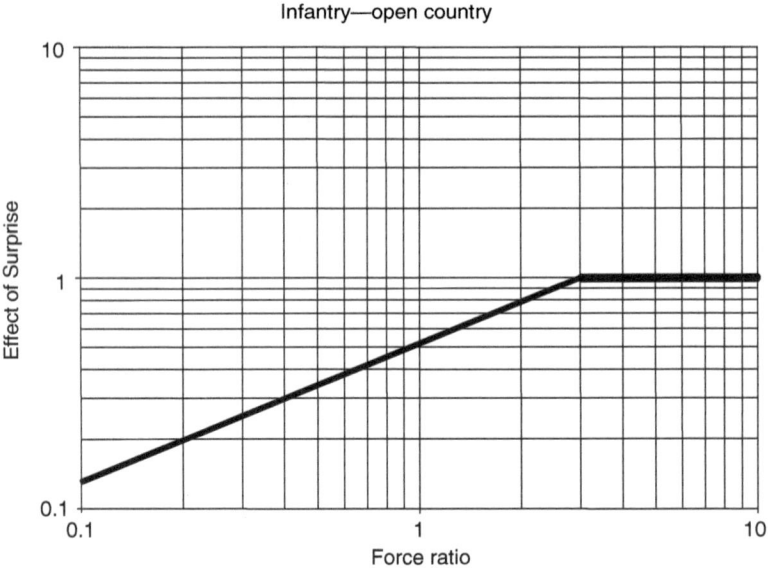

Infantry—open country

quantifying the importance of maintaining an effective balance between weapon numbers and types and how those weapon systems are manned and led. The work has further provided significant insight into our manning ratios and pointed to, for example, the significant risks of decreasing existing ratios of officers and senior noncommissioned officers (SNCOs) to other ranks (ORs). One example of the use of this information is that weapons operated by individuals acting heroically are three times as effective as those without and the relative probability of an individual acting heroically was 0.14 for officers, 0.08 SNCOs, 0.013 ORs. Additionally the work showed that participation in combat was directly related to direct supervision by officers or officers plus SNCOs, while on average a hero gave an improvement in defence effectiveness of a factor of 5.0.

Another issue to emerge from historical analysis is the presence of statistically significant and enduring correlation of measures of performance and effectiveness with nationality. By statistically significant, we mean that the nationality factors are too large and consistent to be attributed to chance. The effects have been found

Figure 2.4 National variations in defence effectiveness

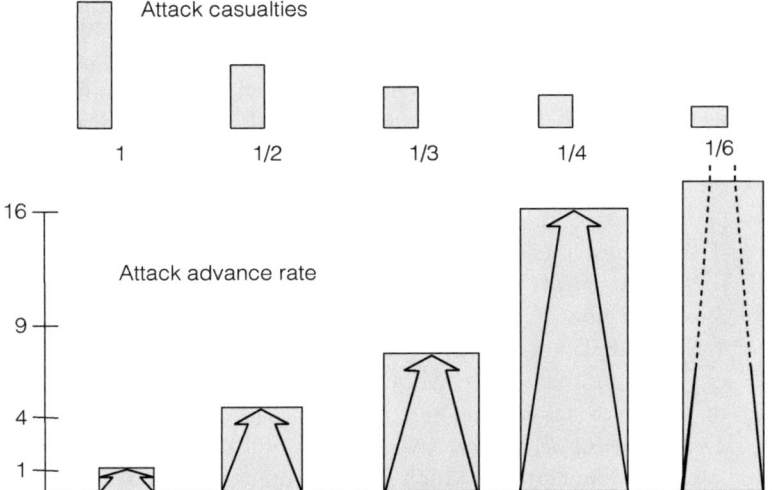

to be consistent over 50–100 years (including very recent conflicts), and in land, sea and air combat. By way of illustration, Figure 2.4 shows the radical differences that were found in the areas of relative attack casualties and relative attack advance rates between five different nationality groupings: at the extreme of differences, the force from the nation on the right suffers only one-sixth of the casualties of the force comprising the nationality of the first group on the left, and can advance at over 16 times the rate—assuming that all other factors are equal. There are clear operational implications for this work, not least in campaign planning and high-level strategic planning.

HUMAN FACTORS IN MODEL DEVELOPMENT

The need for, and use of, historical analysis to assist in model development has already been mentioned. Examples of the models that have been, and are continuing to be, improved range from NEMO, a high-level fast-running war-game tool to CLARION, a new land/air deterministic closed constructive simulation used extensively for high-level studies by CDA, to SIMBAT, a new battle-group model. The last is probably the most sophisticated and complex use of historical analysis in model construction. SIMBAT incorporates the following human factors:

- *Fatigue.* Fatigue affects the surveillance and direct fire process. Too much fatigue forces a unit to rest, although SIMBAT does not model the effect of fatigue on the quality of decisions that a commander makes.
- *Range to trials degradations.* Range to trials degradations, trials to war degradations, and participation, all affect the direct fire process. Range to trials degradations reflect the loss of targeting information that is available on a range shoot, and procedural approaches to casualty reduction.
- *Trials to war degradations.* Trials to war degradations represent the effect of fear on weapons handling.
- *Shock and surprise.* Shock and surprise are modelled in SIMBAT as separate factors. A unit may be shocked and/or surprised, or suffer neither effect. Shock and surprise reduce the ability of a unit to cause casualties during a battle.
- *Overall combat degradation.* An overall combat degradation factor has been developed, which can be used to represent factors such as cohesion or unit training.
- *Participation.* Participation represents the willingness of a component fire unit to fire. Low unit participation reduces the ammunition consumption within the unit.
- *Morale.* The morale check necessary to initiate close combat reflects the willingness of a unit to accept casualties. If a unit is not willing to accept the level of casualties likely from the encounter then it will not close, causing the attack to stall in front of the defence.
- *Disorganisation while reforming.* The disorganisation experienced by a unit that has successfully closed and won close combat has been observed in a number of historic accounts. Until a unit has reformed it is particularly vulnerable to counterattack. The time to reform varies with the time of day: night; dawn/ dusk; or day.
- *Local perception.* The local perception of the tactical situation affects tactical decisions. It represents an important part of what a unit in SIMBAT is willing to do. The better the commander can keep a unit informed of the overall picture the more appropriate a unit's local decisions will be to the overall conduct of the commander's intent. Dependent on local perception, it is possible for a unit to be too cautious, perform reasonably well, or act rashly. These decisions are controlled by user-defined conditions for making a tactical decision.

THE FUTURE

In addition to the types of research and modelling indicated above, our human sciences community is attempting to incorporate combat participation in a model called ModSAF for use in training and synthetic-environment applications. The original model on which ModSAF is based was designed to examine infantry close combat (where individual soldiers are represented and engagements are typically at subunit level), because this level of combat has the most potential for a successful amalgam of historical and psychological data. The model includes the following hypotheses:

- Participation is governed primarily by experience. If a soldier is well trained and has suffered minimal exposure to combat, he will be more likely to engage.
- Physical surprise influences participation. If a soldier is outflanked by an enemy, he is less likely to engage.
- Participation is influenced by a 'weapon power effect'. The more effective a weapon system a soldier has, the more likely he is to engage.
- The participation of individuals is dependent on the degree of participation displayed by other members of their subunit. Soldiers are likely to follow the behaviour of their comrades rather than act in isolation.
- Soldiers in more cohesive subunits (as defined by a psychological rather than doctrinal use of the term 'cohesion') are more prone to radical 'all-or-nothing' shifts in participation, and may therefore be more prone to the effects induced by shock action.

CONCLUSION

This chapter is only an overview of the UK Ministry of Defence initiatives in incorporating human factors into simulation and modelling. The work is very time-consuming. It requires a commitment in resources and investment before worthwhile output is achieved. The reason we have been successful to date in persuading others in the Ministry of Defence to secure funds to pursue this work is that, as we have already stated, it is the right thing to do. And it is in sympathy with our prevailing view of the future battlespace—that is, that conflict is, and will remain, essentially a human activity, in which the human qualities of judgement, self-discipline and courage, the moral component of fighting power, will endure. To out-think, break, and if necessary kill an opponent, while retaining the moral high-ground, will be fundamental, if not essential, to the success of future operations. It is difficult to imagine military operations that will not ultimately be determined through the physical control of people by people. Technology must not be allowed to displace human intent or the decision of a commander. Rather, we need to harness the new information-age way of doing things, ensuring that data do not overcome wisdom in the battlespace. Ultimately, we need to guarantee that real leadership—that which makes men fight when all their instincts tell them otherwise—will act in synergy with the capabilities we design for the future, and not be hindered by them.

3 | Close combat: lessons from the cases of Albert Jacka and Audie Murphy

Michael Evans

He had been to touch the great death, and found that, after all,
it was but the great death and was for others.

Stephen Crane, *The Red Badge of Courage*

WINSTON CHURCHILL ONCE said that danger in war, like good
champagne, should be sipped slowly rather than swallowed quickly.
However, in the history of arms there have always been soldiers
who have swallowed quickly and deeply of danger. In the 20th
century, Albert Jacka of Australia and Audie Murphy of the USA
were two such men. Although separated by differences in nationality and generation, they symbolise the common martial spirit of
the foot soldier of Anglo-Saxon democracy. They are worthy of
serious study because their experiences go some way towards
explaining the phenomenon of personal success in the impersonal
conditions of industrialised war.

This chapter analyses the qualities required for enduring
achievement in close infantry combat. Most individuals are capable
of sudden action or sudden collapse in a crisis, but relatively few
have the psychological capacity to be great combat infantrymen in
war. This kind of soldiering is a rare gift. As Gerald F. Linderman,
the distinguished historian of the American combat experience, has
noted, for most men continuous success in battle is something that,

in terms of psychological and physical effort, cannot be mustered for long periods of time without risking the very destruction of the human self.[1]

This chapter explores five themes. First, it examines the difficulties in reconstructing the world of combat using insight from history, literature and philosophy. Second, an outline of the similarities in background between Albert Jacka and Audie Murphy is provided. Third, case studies of Jacka and Murphy as exemplars of the art of close combat are presented, along with some explanations for their extraordinary success. Finally, an attempt is made to deduce some enduring lessons from the soldiering of both men.

PROBLEMS IN RECONSTRUCTING CLOSE COMBAT

The American combat historian, Brigadier General S.L.A. Marshall, once wrote that the battlefield is the epitome of war and those that study it must remember 'the tactical fact which is at once the simplest and most complex in the military art—man himself as a figure on the field of combat'.[2] The tangibles of war, command, logistics, operations and strategy, describe a sequence of routines by which war can be planned in battles and campaigns. Such routines represent war at the macro level; but they are its skeleton rather than its flesh.

The flesh of war is the fighting man. Unlike senior officers planning the broad sweep of battle at headquarters, ordinary combat soldiers operate on the line—a narrow, micro-level world of immediacy and confusion. It is the ordinary soldier who experiences the transformation of higher strategy into actual tactics and, with it, the fragmentation of war from planned battle into unpredictable combat. What is friction to the commander is chaos to the combatant. In action, most soldiers see only one angle of the battlefield. When action is over they do not usually write the official reports that go up the line, and as the reports go higher and further to the rear so they move away from the reality of the front. For these reasons, the soldier's war has been aptly described as 'the great secret of military history'.[3]

The paradox of those 'doing the fighting but not doing the writing', the contrast between the tangibles of battle developed at headquarters, and the intangible features of combat on the line make the reconstruction of military action one of the most challenging of historical exercises. The British commander General

Sir Ian Hamilton captured the nature of this challenge when he observed, 'on the actual day of the battle naked truths may be picked up for the asking: by the following morning they have already begun to get into their uniforms'.[4]

For many soldiers, combat takes place in an interior psychological world of terror, confusion and violence. Consequently, combat is sometimes described as being incommunicable—a personal experience which compresses the greatest opposites into the smallest space and the shortest time, often detaching memory or suppressing it. To reconstruct combat, one has to overcome its many apparently incommunicable obstacles. These obstacles include the problem of imperfect and painful recollection, the presence of biased or contradictory evidence, and the distorted memories of veterans disconnected in time and space, some sharp, some dim, some in and some out of focus. The American writer James Jones put it well in his great war novel *The Thin Red Line,* when he wrote of the surviving veterans of the Guadalcanal campaign, 'one day one of their number would write a book about all of this, but none of them would believe it, because none of them would remember it that way'.[5]

Perhaps the most formidable problem facing historians is that combat soldiers inhabit what J. Glenn Gray, the American soldier–philosopher, has called 'the tyranny of the present'.[6] In this environment, past and future, reflection and reasoning, are almost alien concepts. Modern infantry combat is rooted in a psychology of crisis in which sensory details dominate and space and time are compressed. As Gray puts it, 'the great God Mars tries to blind us when we enter his realm, and when we leave he gives us a generous cup of the waters of Lethe to drink'.[7]

Experienced front-line soldiers are forced into 'the tyranny of the present' because they know instinctively that they possess only a finite amount of courage and endurance and that their lives are always at risk. Combat infantrymen have to carry a terrible burden and it is this: no matter how perfect a man may be as a soldier, he has no guarantee that he will not ultimately be killed. Soldiering is the only human profession in which success is shadowed by imminent physical destruction. It is utterly different from anything found in modern civil society. In the civilian world, no-one is asked to die in the name of business cost-effectiveness or sporting excellence, and civilian celebrity is not the same as military heroism. On the battlefield, a random explosion or a roving sniper might end days, months, and even years of brilliant soldiering. Roger Spiller puts it well by observing,

> when a soldier moves forward against fire, he steps beyond the boundaries of anything we understand. Then, centuries of military science are at the mercy of one bullet, and if reason is at play it must expend its power in forms so different that they have eluded us thus far.[8]

A combat soldier, then, inhabits the realm of the present by necessity; he is forever at the whim of chance, a constant fugitive from the law of averages. Combat veterans know the truth of the old saying 'a battlefield is at once the playroom of all the gods and the dancehall of all the furies'.[9] It is not surprising, then, that so many combat veterans remain silent about the harrowing experience of military action. The angel of death sits on the shoulder of every soldier in the field, and ultimately every soldier knows he is present. Those veterans that survive combat know they have avoided becoming a victim of the charge of that most persistent member of the Four Horsemen of the Apocalypse—the pale rider called War.

Because combat experience tends to hide in the minds of men, it is perhaps the most difficult world for scholars, especially historians, to enter. This is why novelists have often succeeded where historians have failed. The novelist can fictionalise reality, hide identities, obscure pain and bury memory. The scholar cannot do this. The enemy of the scholars of war is the combat soldiers' imprisonment in a world that is often unimaginable to the civilian—a world of repetitive danger in which the suspension of time and the freezing of emotional reaction confront all of society's romantic ideals about masculinity and heroism. The great casualties of this confrontation between reality and idealisation are written record and memory, the basic tools of the historian.

Outside fiction, there have been only a handful of modern writers who have probed seriously the human characteristics of combat. They include the 19th century French military theorist Colonel Ardant du Picq, whose pioneering book, *Battle Studies*, was first published in 1880; S.L.A. Marshall's 1947 classic work, *Men Against Fire*; and J. Glenn Gray's remarkable 1959 study, *The Warriors*.[10] Although these writers are separated by different eras and different approaches, broadly speaking their work highlights three common themes: the social nature of war; the enduring chaos of fighting; and the importance of understanding the human dimension in battle.

All three writers view the experience of battle as social and not technological in nature. They emphasise not the predominance of weaponry but the critical factor of human cohesion in combat. As Gray put it, 'loyalty to the group is the essence of fighting morale'.[11] Men do not usually fight as duellists. They bond together in a disciplined formation, because it is in the formation that a man's natural fear and aversion to fighting is counterbalanced by the will of the primary group. This is why, wrote du Picq, the Greeks and Romans fought in phalanxes and legions.[12]

Du Picq, Marshall and Gray all recognised the chaos of war and the threat this posed to tactical unity and unit cohesion. They understood that combat is not unitary and symmetrical but fragmentary and asymmetrical. The battlefield, and now the modern battlespace, is quickly fractionalised by firepower. Unpredictable movement and confusion often dissolve unit formation. Combat resembles an eddying tide, not a single incoming wave. As Marshall noted, 'in combat almost nothing has the appearance of juncture and of hanging together. Viewed from above, an attack would appear not unlike the disparate movements of a colony of water bugs'.[13]

All three writers also emphasised the profoundly human activity of war. Men may wear the same uniform but they are never *uniform* in their reaction to combat or in their feelings of closeness to death.[14] Gray noted the existence of various subspecies of front-line men defined by their attitude towards death. There are, he argued, soldier–adventurers, to whom action and experience are everything in war. There are soldier–killers, deadly efficient men often devoid of remorse or reflection.[15] There are otherworldly soldiers, governed by spirituality; and there are serious professionals, to whom confronting death is part of their sworn duty.[16] This is not, of course, a definitive list, but it gives an indication of the human differences that are often clothed by a common uniform.

We can glean some knowledge about the hidden and complex world of front-line infantry warfare from Australia's Albert Jacka and America's Audie Murphy. By examining briefly the military careers of these two outstanding soldiers, we can measure their personality traits and military performance against several of the propositions advanced by du Picq, Marshall and Gray in their efforts to explain the nature of close combat.

SIMILARITIES BETWEEN ALBERT JACKA AND AUDIE MURPHY

Albert Jacka is often seen as personifying the Anzac spirit. The official historian C.E.W. Bean hailed him as 'Australia's greatest fighting soldier'. Jacka is probably the most famous of all Diggers— a man idolised by his comrades, decorated by the authorities and mythologised by the popular press.[17] His picture was widely used in recruiting posters in Australia during World War I. Prime Minister Billy Hughes considered Jacka's fame to be so great that he offered, unsuccessfully, to take him out of combat in return for political support of conscription. When Jacka died in January 1932, Melbourne came to a halt, and more people attended his funeral than that held in 1931 for General Sir John Monash.

Audie Murphy remains the most decorated soldier in American military history and is probably the most famous of all GIs. In 1964, the distinguished combat soldier and Congressional Medal of Honour winner Major General Keith Ware described Murphy as 'the finest soldier I have ever seen in my entire military career'.[18] Audie Murphy died in 1971 and is buried at Arlington National Cemetery, where, along with the Tomb of the Unknown Soldiers and the graves of the Kennedy brothers, his resting place is among the most visited sites.[19]

There are many similarities between these two celebrated warriors separated by two wars. Both were short and physically unimposing; both were outwardly uncomplicated characters; both were volunteer soldiers; both were country labourers—Jacka was from rural Victoria and Murphy from rural Texas. Both were teetotallers and non-smokers; both were undemonstrative, with a dislike of ceremony; both had limited formal education but each possessed an intuitive intelligence. Both men received wartime commissions in recognition of their tactical skills; both seemed to find in combat a purpose and a sense of worth that they were unable to discover in civilian life; and neither man was particularly religious. Both suffered illness due to their wartime experiences. In postwar society both men died young—Jacka at 39 and Murphy at 46; and both died bankrupt and disillusioned.

THE CASE OF ALBERT JACKA

Albert Jacka became a national hero when he became the first man in the 1st Australian Imperial Force (AIF) to win a Victoria Cross in World War I. Serving with the 14th Battalion, 4th Brigade at Courtney's Post, Monash Valley on Gallipoli in May 1915, Jacka led a small-unit counterattack that killed seven Turks and recaptured a company trench. His quip to his commanding officer, 'Well, I managed to get the beggars, Sir', made him famous in households throughout Australia.[20] After Gallipoli, Jacka was sent to officer-training school, where he topped the class in the tactics course. Once commissioned as a second lieutenant Jacka was sent to France, where his exploits continued on the Western Front. At Pozières Ridge in 1916, Jacka was hit seven times leading a counterattack through a hail of bullets on a German position. He killed a dozen Germans with revolver and bayonet and captured 50 more. The ferocity of Jacka's action astonished onlookers. Charles Bean called the counterattack 'the most dramatic and effective act of individual audacity in the history of the AIF'.[21] For this deed Jacka won the Military Cross. At Bullecourt in April 1917, Jacka again distinguished himself in small-unit action and received a Bar to his MC. In May 1918 at Villers-Bretonneux he was badly gassed, and never again returned to the front.[22]

What was the secret of Jacka as a combat infantryman? Over 20 years ago, the University of Melbourne historian Lloyd Robson described Jacka as an enigma—clearly a hero figure, but one whose frequency of bravery probably harboured a death wish.[23] Jacka's combat motivation is open to other interpretations. It is equally possible that he was a soldier–adventurer as described by J. Glenn Gray. For soldier–adventurers, wrote Gray, 'battle appears to be their very element, and in that element men will not hesitate to pay them homage'.[24] Battle was Jacka's element: he energised combat. As Corporal Charles Smith recalled, Jacka seemed strangely suited to the grim surroundings of the trenches, and was held in awe for his skills by his comrades.[25]

Jacka was confident, intuitive and audacious in action. Despite suffering serious gunshot wounds, shellshock and gassing, his thirst for combat was such that he would often patrol alone in no-man's-land.[26] When asked by a senior officer why he had led an apparently suicidal assault at Pozières Ridge, Jacka replied, 'I was a VC—what else could I do?'[27] Clearly Jacka was an inspirational soldier, the main prop of the 14th Battalion, which became known throughout the AIF as 'Jacka's Mob'. Brigadier General Charles Brand, commander of the 4th Brigade, once commented, 'a company under his [Jacka's] lead was as good as an extra battalion to me'.[28]

Out of combat, Jacka seems to have been a difficult personality. He was abrasive, ambitious for promotion and expansive of his combat exploits. C.E.W. Bean said that Jacka was 'a good fellow of a rather crude type'—a remark that perhaps captures social prejudice towards a working-class man with an officer's commission.[29] When he was not awarded more decorations and not promoted beyond captain, Jacka became openly contemptuous of many of 'the heads' or senior brigade officers.[30] They in turn regarded him as a superb fighter but portrayed him as a poor officer.[31]

To understand Jacka more fully, we need perhaps to appreciate what J. Glenn Gray has written about the appeal of battle. Gray was a US counterintelligence officer in World War II, and his observations about combat can be found in his book *The Warriors*. He argued that combat has a three-fold appeal: first, the delight in war as a *spectacle*; second, the delight in *comradeship*; third, the delight of dealing out *destruction*. Interwoven with these elements of spectacle, comradeship and destruction, Gray identified a fourth element that pervades the first three—that of a delight in *danger*.[32] As Gray put it, battle can be 'the one great lyric passage' in a man's life.[33] It is significant that the great American war correspondent Ernie Pyle, who was killed on Okinawa, came to a similar conclusion on the appeal of combat. 'War', he wrote, 'is vastly exhilarating . . . There is an intoxication about battle, and ordinary men can sometimes soar clear out of themselves on the wine of danger–emotion.'[34]

Evidence suggests that Jacka tended to fit Gray's categories while soaring on Pyle's wine of danger–emotion. He thrived on the spectacle of war; he excelled at soldiering at the head of 'Jacka's Mob'; he proved extraordinarily skilled at destroying Turks and Germans; and he relished the thrill of danger. We get a clue to Jacka's love of fighting from his contemporaries. Captain E.J. Rule, author of *Jacka's Mob*, one of the best Australian memoirs of World War I, wrote: 'Bert Jacka did not stumble into fame . . . There were, in fact, few men in fact who revelled more than he in the intoxication of success, for he was very human, though undemonstrative.'[35] In an interview in 1977 one of his old comrades, Percy Bland, described Jacka as a 'a fanatic, a soldier fanatic. He could just not help attacking'.[36] In 1989, Jacka's biographer, Ian Grant, concluded that his subject was 'compulsorily drawn to battle because he found he excelled at it'.[37]

How does a soldier–adventurer meet normal life after such wartime fame? Jacka returned to Melbourne in October 1919 to a Roman triumph. His biographer notes, 'Jacka's return became the closest re-enactment that Australia was ever to see of a Caesar returning victoriously from the field of battle.'[38] A convoy of 85 cars headed by the Governor-General escorted 'the symbol of the spirit of the Anzacs' into a cheering city, where Jacka was mobbed by adoring citizens.

In the 1920s Jacka remained in the public eye. He went into business, married and eventually became Mayor of St Kilda. But both his business and marriage faltered with the onset of the Great Depression. We know relatively little about Jacka's medical history after the war but we do know he suffered from ill-health. A fellow Victoria Cross winner, Donovan Joynt, remarked that Jacka's hair went from black to grey in a matter of months during the economic crisis of 1931.[39] On his death in January 1932 at age 39 from kidney disease, Jacka was a burnt-out shell who looked nearer 60 years of age.[40] He was widely seen to have been an economic martyr, a victim of another kind of war—the Great Depression—that he could not beat. Jacka, the soldier–adventurer, exemplar of combat, the man who found lyricism in fighting, the symbol of the Anzacs, probably never truly adjusted to civilian life, and eventually its complexities helped to kill him.

THE CASE OF AUDIE MURPHY

If Jacka was an outstanding soldier–adventurer, a combat practitioner who sought recognition for great deeds, Audie Murphy probably best fits the profile of what J. Glenn Gray has identified as the soldier–killer. Murphy, born in Texas in 1924, is the most decorated soldier in the history of American arms. His background as the son of a sharecropper was even more impoverished than that of Jacka. Short, skinny and baby-faced, Murphy was turned down by both the US Marines and the paratroops as unsuitable for combat duty.

Eventually Murphy joined the US Army's 3rd Infantry Division, where he was initially nicknamed 'Baby'. He served in Italy, France and Germany and became the Army's most famous combat soldier, finishing the war with 29 medals—including every known American decoration for valour—and a personal tally of 241 enemy dead.[41] Nothing seemed beyond him as an infantryman: he destroyed machine-gun nests, killed snipers and wiped out enemy patrols; he was at home in urban and rural actions.

Murphy's most famous exploit came in January 1945 in the snow of the Colmar Pocket in south-eastern France, when he was servingas a company commander in the 1st Battalion, 15th Infantry Regiment, 3rd Infantry Division. Two reinforced German rifle companies, a force of some 250 infantry, and six Tiger tanks attacked his battalion's right flank and threatened to overrun it. First Lieutenant Murphy climbed aboard a burning tank destroyer and began to fire the .50 calibre machine gun while simultaneously directing artillery fire by field telephone. In about 30 minutes he killed 50 German soldiers, repelling the enemy assault. For this extraordinary personal action he was awarded the Congressional Medal of Honour.[42]

What was Murphy's secret? Unlike Jacka, who was apparently concerned with military status and decorations, Murphy developed a cynical fatalism about war. As he once put it, every dead soldier won a medal—the Wooden Cross.[43] His autobiography *To Hell and Back,* first published in 1949 and still in print today, is a remarkable study of the infinite small threats that have to be overcome by men on the front line.[44] In his book, Murphy argued that 'coolness and calm fury' were the two greatest attributes in successful infantry combat; it was more important to be audacious than courageous because audacity, unlike courage, was a tactical weapon.[45] Here there is great resemblance to Jacka's approach to combat. In situations of danger, recalled Murphy, 'things seem to slow down for me . . . Things become very clarified'.[46] It is clear that Murphy was a cool, deadly efficient soldier or, as Ernie Pyle wrote, 'a senior partner in the institution of death'.[47]

After the war, like Jacka, Murphy became a national hero. He was feted by a quarter of a million people on his return to San Antonio, Texas, the home of the Alamo. The media packaged him as a symbol of pure Americana and Norman Rockwell goodness, the baby-faced kid from Texas who had taken on the cream of the German Army and emerged victorious. The weight of his fame denied him even a routine psychological debriefing by the US Army. With no preparation he was dropped into the mainstream of American civilian life. He went to Hollywood and became a film star, making 44 movies. These were mainly low-budget Westerns, which established Murphy 'as *the* B-Western star of the post-war period'.[48] However, he occasionally made major films, notably a fine version of Stephen Crane's *The Red Badge of Courage,* directed by John Huston in 1951. In 1955 Murphy, in what may be a unique decision by a combat veteran, decided to play himself in a major movie version of *To Hell and Back*. The film became a huge international box office success, and has been listed as one of the 10 best American movies of World War II.[49]

For a short period during the 1950s, Audie Murphy was one of the most popular film stars in the English-speaking world.[50] Like John Wayne in *Sands of Iwo Jima*, Audie Murphy in *To Hell and Back* had a notable cultural influence on a generation that was later confronted by war in Vietnam. In 1976 the Vietnam veteran and political activist Ron Kovic recalled Murphy's impact:

> I'll never forget Audie Murphy in *To Hell and Back*. At the end [of the film] he jumps on top of a flaming tank that's just about to explode and grabs the machine gun blasting it into the German lines. He was so brave I had chills running up and down my back wishing it were me up there. There were gasoline flames roaring around his legs, but he just kept firing that machine gun. It was the greatest movie I ever saw in my life.[51]

Yet, despite its success and influence, Murphy privately despised Hollywood's treatment of his autobiography as a 'Western in uniform'.[52]

How does a soldier–killer survive in peacetime? Because of the abundant documentation of Murphy's postwar Hollywood career we know much more about him than we do about Jacka. It is clear that Murphy was unable to put the war to one side. His civilian life was little more than a postscript to infantry combat. Behind his wartime fame and Hollywood glamour, Murphy had a damaged psyche. He lived in a world of nocturnal paranoia in which insomnia and nightmares alternated; where hypersensitivity to noise made him twitch like a cat and flashbacks to combat made him freeze up like a zombie.

Frequently, he could sleep only with lights on, with a pistol or revolver under his pillow and with the aid of prescription drugs. His first marriage failed, partly because Murphy was plagued by violent nightmares from which he sometimes woke to fire a revolver or pistol around the bedroom, shattering mirrors, clocks and light bulbs.[53] An obsession with firearms, notes one of his biographers, 'became the central prop in the internal theater of his psyche'.[54] Murphy had all the classic symptoms of what medical science knows today as post-traumatic stress disorder—a sense of morbid futility, isolation, sleeping disorders, restlessness and hair-trigger anger. In the European theatre in 1944, 26 per cent of US casualties were neuropsychiatric.[55] Audie Murphy was one of the undiagnosed—a loner with a sundered spirit.

Through movie acting, Murphy sought a life of surrogate action as the good hero. It did not work. He was the exact opposite of John Wayne, whom the American cultural historian Garry Wills has called 'the warless "war hero"'—because Wayne thrived on surrogate action but never served a day in the military.[56] Murphy, an authentic war hero, was bored by movie acting; he disliked actors and hated the artificiality of Hollywood life. He steadily withdrew into a self-destructive existence—this time on the dark edge of the underworld. He carried a Derringer, associated with gangsters and racketeers, and became a compulsive gambler and womaniser.[57] When Murphy was killed in May 1971 at the age of 46 in an air crash, he was bankrupt and his movie career had collapsed.

What marks Audie Murphy as a soldier–killer? Repetition, not progression, is the key to his identity—an identity shaped by infantry combat. J. Glenn Gray provides a description that seems to fit Murphy. The soldier–killer, he writes, is a restless man for whom there is no accessibility to normal satisfaction and no peace of mind:

> [Soldier–killers] rarely can feel remorse, they experience no purgation and cannot grow . . . Though there may be a fierce pride in the numbers destroyed and in their reputation for proficiency, soldier–killers usually experience an ineffable sameness and boredom in their lives. The restlessness of such men . . . is notorious.[58]

These features of boredom and restlessness are reflected in Murphy's book, *To Hell and Back*. After killing for the first time Murphy writes: 'I feel no qualms; no pride; no remorse. There is only a weary indifference that will follow me throughout the war.'[59] Elsewhere in the book, he speaks of how combat action served to 'spike the rot of existence' and how courage was born from 'the utter boredom of static warfare [that] drives men to strange deeds'.[60] These feelings continued in peacetime. In 1967 Murphy described himself as 'an ancient young man' whose 'fight for a long time [has been] to keep from being bored to death'.[61]

Many in Hollywood who worked with Murphy recall his dangerous intensity and his soldier–killer eyes. He frightened the film actor Kirk Douglas, who described him as 'a vicious guy'.[62] Film directors John Huston and Don Siegel referred to Murphy respectively as 'the little killer' and 'the killer of all time'.[63] In 1997 Denver Pyle, a veteran US film actor who made four films with Murphy, recalled: 'You could look [Audie] in the eye and know that he would think nothing of killing you . . . you can see it in his [film] work . . . He was deadly.'[64] In some of Murphy's Hollywood action scenes, directors avoided camera close-ups of the Texan's face for fear of frightening the audience.[65] Significantly, one of Murphy's best and most revealing film performances was as a bored killer in the 1959 Western *No Name on the Bullet*. In nearly every scene in this film, Murphy oozes sulky menace—from the movement of his cold eyes to the terse dialogue that he delivers with flick-knife intensity.

LESSONS FROM THE COMBAT CAREERS OF JACKA AND MURPHY

What lessons can we glean from the unusual careers of Albert Jacka and Audie Murphy? Three come to mind. First, both men reaffirm the eternal truth, validated by three thousand years of human conflict, that success in war usually requires first-class infantry to close with, and to finish off, an enemy. There is no other way of winning—if by winning we mean the seizure of ground, the control of resources and the political defeat of an enemy. This was accomplished in the world wars and more recently in Kuwait. Good combat infantry do not spring fully armed and trained out of the earth, as they do in Greek mythology. They are made and moulded into the special fraternity of fighting men—of whom Jacka and Murphy are two outstanding examples.

Second, Jacka and Murphy, albeit in different ways, remind us of the strange, almost mystical appeal of close-quarter combat. Many young men at some point in their lives contemplate the famous choice of Achilles—to risk all, perhaps to die young, but after having lived gloriously; or to live long, securely but inconspicuously. We need to remember that close combat, for all its inhumanity and horror, has a long history of eager participants.[66] In Homer's *Iliad*, a young Greek warrior asks his companion why they are risking their lives in battle. The companion explains: because it is the fate of man to die, he can choose war to confront death personally; to stalk the great stalker in the ultimate game of risk.[67] It was the Roman soldier–philosopher Marcus Aurelius who observed that an 'effective aid to attaining contempt of death is to review in your thoughts those who have clung tenaciously to life'.[68]

Combat soldiers may wear uniform, but they are never uniform in their reaction to combat. In modern societies there are no natural soldiers. No-one who knew Jacka in 1914 or Murphy in

1941 could have foretold that these obscure country boys would become two of the greatest combat infantrymen in the military history of the 20th century. We must always remember, therefore, that for every Albert Jacka there is a Henry Murray. Murray, a Tasmanian, was the most highly decorated soldier of all the British Empire armies during World War I. In peacetime, he led an exemplary life as a grazier and militia commander in Queensland.[69] For every Audie Murphy there is a Maurice Britt. Britt, an Arkansas football star who lost an arm at Anzio, was the second most decorated American soldier of World War II. In peacetime, he pursued a successful career in Arkansas politics.[70] Murray and Britt demonstrate that military heroes can survive the harrowing experience of combat and go on to prosper in peacetime.

Yet it has been the enigmatic and flawed personalities of Jacka and Murphy, not the well-adjusted characters of Murray and Britt, that have captured the public imagination in Australia and the USA. This leads on to a third lesson: Jacka and Murphy are proof that it is men-at-arms, not military machines, that give Western society the mythology to nourish its martial pride and traditions. As the American philosopher Paul Zweig puts it, 'the warrior's duel with death provides culture with its essential tools, its founding myths, its knowledge of the world'.[71]

We remember Achilles in the *Iliad* not because of his golden armour and his sleek chariot but because of his flawed character and personal fighting prowess. Similarly, we honour Jacka and Murphy for the way they faced the lethal technology of the industrial battlefield. Their achievements represent a triumph of the spirit by ordinary humans who, in the maelstrom of war, proved capable of undertaking extraordinary feats of valour. Both individuals showed that success in combat is victory over fear, not the absence of fear.

For these reasons, Albert Jacka and Audie Murphy will continue to be remembered and honoured in the societies from which they sprang. Jacka remains the archetypal Anzac hero in Australia. In 1950 there was a furore when an Anzac Day radio broadcast by the Australian Broadcasting Commission (ABC) portrayed Jacka as an illiterate wood-chopper, whose bravery was that of an unthinking yokel. The Jacka family was outraged and the matter was raised in the federal parliament. In the House of Representatives, Allan McDonald, the member for Corangamite in Victoria and a comrade of Jacka's in the 14th Battalion, rebuked the ABC for 'defaming a dead soldier and patriot'.[72] He went on to state: 'Those of us who knew him [Jacka] best and loved him most will always remember him as an incomparable soldier, a tried and trusted leader, and a loyal and faithful friend . . . One of the greatest fighting soldiers that Australia has ever produced.'[73] According to McDonald, when Jacka was in the battle line each member of the 14th 'became, to the top of his capacity, himself an Albert Jacka'.[74] In his 1992 autobiography, the soldier–politician Henry 'Jo' Gullett referred to McDonald's speech as 'the most moving and effective speech I heard in my years in Parliament.'[75]

In 1977 Jacka was the subject of a perceptive documentary film, *Jacka VC*, which included interviews with many of his former comrades.[76] In 1989 the first biography of Jacka appeared. Its publication coincided with renewed national interest in Anzac Day, and came only two years after a symbolic 'welcome home' march for Australia's neglected Vietnam veterans.[77] In 1999, the 40th anniversary edition of *Time Australia* listed Albert Jacka as one of its 'Southern Stars'—one of the 10 most influential military figures produced by the South Pacific region in the 20th century.[78] Jacka appeared alongside such war heroes as fellow Australians John Simpson Kirkpatrick, Arthur Roden Cutler, Ernest Edward 'Weary' Dunlop; the New Zealanders Nancy Wake and Charles Upham; and the Papuan and New Guinean 'Fuzzy Wuzzy Angels' of the Kokoda Trail.[79] In Melbourne, the anniversary of Jacka's death is still commemorated publicly each year at his graveside in the St Kilda cemetery.

In the USA and abroad, Audie Murphy continues to be revered by many people as the bravest of all GIs. *To Hell and Back* remains in print and Murphy's career continues to attract both historical and literary interest from many parts of the world.[80] The US Army has not forgotten its most decorated soldier. Since 1989 the Army has had a Sergeant Audie Murphy Club for outstanding members of that rank. During 1996 Audie Murphy was inducted into the National Cowboy Hall of Fame and the Audie Murphy Research Foundation was set up in Santa Clarita, California, to 'collect, preserve and make available to the public historical information concerning the life and times of Audie Murphy'.[81] Since 1997 there has been an Audie L. Murphy—American Hero Commemorative Stamp Campaign. Nostalgia and admiration for the passing of the World War II generation have fuelled much of the reverential interest in Murphy in recent years. In the newsletters of the Audie Murphy Research Foundation, comments to the effect that Murphy personified 'the last American hero' are commonplace.[82]

CONCLUSION

The movement of men on the field of battle to close with an enemy, to kill or to be killed, whether from phalanx or foxhole, is the oldest instrument of statecraft. Close combat has been the defining feature of war in the past and is likely to remain so in the future. As the classical scholars Victor Davis Hanson and John Heath remind us, 'the entire freight of Western civilisation—constitutional government, individual freedom, capitalism, Christianity—has spread through the blood and iron of Western infantry'.[83]

The careers of Albert Jacka and Audie Murphy are melancholy testimony to the great debt we owe to the footsoldiers who helped preserve Western civilisation in this century. Modern Western democracies are now allegedly post-heroic and seem to have become averse to suffering military casualties. Since the 1990–91 Gulf War it has often been argued that Western societies should fight future wars at long distance with air-delivered precision munitions and so seek to avoid the human costs of close combat.[84] Yet there is no shortage of dangerous adversaries around the world who will field armies, militias and guerrillas to exploit future Western unwillingness to deploy ground forces.

Yet, reliance on stand-off information-age technology alone does not offer a permanent solution to managing intractable human conflicts—nor is it likely to change the horror of battle or the reality of death. To meet the challenge of war in the 21st century the West will still need that most ancient and trusted of weapons systems—the infantryman. What was true for the Greek hoplite and the Roman legionary will be true for the Australian Digger and the American GI of the 21st century. War will remain a clash of human wills. As Roger Spiller has noted, war is most human at the very place where humanity is in the greatest danger of extinction: in the killing grounds and zones of combat, where men devote every impulse of their mental and physical energy to destroying one another.[85]

Finally, the French military theorist Marshal Maurice de Saxe once wrote, 'War is a science covered with shadows in whose obscurity one cannot move with an assured step.'[86] Insofar as it seems possible, Albert Jacka and Audie Murphy moved through the shadows of war with an assurance of step that is given to few men. In peacetime, each man paid a heavy price for this gift. As we enter the 21st century, Jacka and Murphy remain enduring testimony to the truth that human combat cannot be reduced to technical dynamics—that the strength of a democracy lies not in its material resources, its technology or its economic strength but in the spirit and valour of its citizens.

4 | The last casualty? Public perceptions of bearable cost in a democracy

Hugh Smith

GREAT POWERS AIN'T WHAT THEY USED TO BE

Napoleon once boasted to Metternich that he could afford 10 000 soldiers a month in his campaigns. Conscription and patriotism had together opened new possibilities for military sacrifice. A century later in the Great War the loss of 10 000 soldiers in a single day was stoically accepted. By World War II, profligacy with human lives was less tolerable and less in evidence, at least on the part of the Western powers with regard to their own forces. Yet support for the war remained high, despite heavy casualties.

The war in Korea was rather different. As casualties rose and military stalemate set in, popular support fell away and the USA felt compelled to negotiate a truce. America's experience in the Vietnam War seemed to mark a further turning point. The shift in public opinion against the war, especially after 1968, was followed by a decision to abandon the military effort altogether. The loss of nearly 50 000 military personnel seemed too much for the American community to bear, even when fighting the old and powerful enemy of communism.

Subsequent events reinforced the perception that the USA had lost the willingness to sacrifice its young men (and women) in protecting and promoting its interests. The Gulf War did so in an indirect fashion. Saddam Hussein told the US ambassador before

Iraq's invasion of Kuwait: 'Yours is a society which cannot accept ten thousand dead in one battle.'[1] The Iraqi leader may have been right, as only 148 American soldiers were killed (35 from friendly fire) in comprehensively defeating what was touted as the fourth-largest army in the world. The message was that the USA could win wars with consummate ease and minimal casualties.

Not only were casualties on one's own side less acceptable, there appeared to be growing concern that military action should not kill enemy civilians. Witness the widespread criticism of the deaths in the attack on the Amiriya bomb shelter during the Gulf War. Even enemy soldiers were to be spared. President Bush's halt to the destruction of the Iraqi army in retreat on the road to Basra was interpreted by some as an unaccustomed, even unmilitary, squeamishness on the part of the USA.

US interventions in the post–Cold War era reinforced the perception of decreased tolerance for casualties in more direct fashion. The loss of 18 soldiers in a single incident in Somalia in October 1993—with television showing the body of one of the dead being dragged through the streets—appeared to cause the USA to lose heart and want out. The retreat of the USS *Harlan County* from Port-au-Prince some two weeks later confirmed the impression—all the more so as thugs on the dockside employed by the Haitian regime shouted: 'We are going to turn this into another Somalia.'[2]

America's prolonged hesitation in the 1990s over sending ground forces to Bosnia seemed final proof that it was no longer willing to put its soldiers at risk. NATO's bombing campaign against Yugoslavia over Kosovo in 1999 reinforced the view that the USA was prepared to fight only from a distance. There was no eagerness on the part of America's leaders to prepare for a ground war should the air campaign fail.

Other democracies seemed to have caught the disease. Western European countries were clearly reluctant to commit forces to Rwanda, a faraway country of which they knew nothing. The kidnap and murder of 11 Belgian paratroopers in Rwanda in April 1994 led to strong demands to bring the contingent home. Even Israel was compelled to retreat from Lebanon, in part by public hostility to the continued loss of life. Britain's tolerance of over 600 military deaths in Northern Ireland is no real exception, as the province is part of the homeland.

Since 1945 Japan and Germany have been in a condition less of 'combat reluctance' than of 'combat disqualification'.[3] Constitutional restraints on sending forces overseas, strongly backed by public opinion, made both countries highly anxious about possible reaction to casualties in the event of such commitments. When Japan first sent troops and police to a peacekeeping operation (Cambodia), extraordinary steps were taken to avoid loss of life for fear of domestic criticism.[4]

Even the Soviet Union in the 1980s appeared troubled by mounting casualties in Afghanistan as resentful families spread information about relatives returning in coffins. In countries with a free media, public anxiety can be fed even more effectively.[5] Extraordinary technical advances in the speed, comprehensiveness and range of media coverage of wars have occurred in the past 100 years—from written despatches to photographs to newsreel film to live television coverage with instant analysis by experts (and non-experts).

The media take great interest in casualties, especially those in certain categories, such as conscripts, blacks, females or single parents. In the Gulf War much publicity was given to the despatch of hospital ships and preparations for casualties. Typical, too, was the cover of *Time* in November 1995, which showed a US soldier with the caption: 'Is Bosnia worth dying for?' The media personalise war in many ways: witness the intense interest in the fate of a single US airman, Captain Scott O'Grady, shot down over the former Yugoslavia in June 1994.

The media may be giving the public what they want, but there are deeper changes. The media are increasingly driven by the need for visual images, which in turn influences the message:

> Television, after all, is very good at covering the death of one
> US soldier. But it is far less good at explaining, for instance, the
> many civilian deaths that may have been prevented in central
> Bosnia by UNPROFOR's deployment.[6]

In addition, the specialist defence reporter has all but disappeared, leaving the field to generalists who prefer to focus on 'human interest' stories rather than strategy. Deeper analysis of political and strategic issues thus struggles to attract the attention of both producers and the general public.

In the past, governments were sometimes able to restrain their 'own' media from covering the miseries of war by direct controls or by appeals to patriotism. In World War II *Life* magazine did not publish pictures of American dead until 1943, and no photographs showed dismembered Americans.[7] During the Vietnam War one issue of *Life* published the photographs of all those killed in the previous week, a move that some saw as a turning point in public attitudes.[8] Today, if a war is not fully covered by a nation's media, it will probably be reported by those from other countries or by international rather than national media.

Film and video offer vividness and immediacy—a front seat for the front line—which cannot but affect the minds of a population engaged in war. The speed of reporting, moreover, means that CNN's report today or tomorrow's newspaper headlines are often the first to announce casualties. Politicians must respond as best they can to the spin put on by the media. As early as 1970, the British television commentator Robin Day drew the conclusion: 'One wonders if in future a democracy that has uninhibited television in every home will ever be able to fight a war, however just.'[9]

The revolution in media affairs—another revolution in military affairs (RMA)—lends support to the argument for growing aversion to casualties, and technology promises to spread it even further. As TV and radio programs are beamed down from satellites or reports are clicked around the Internet, even authoritarian governments may find control over the reporting of wars more difficult.

WHAT IS AVERSION TO CASUALTIES?

The argument that democracies have lost the stomach for casualties is plausible, but is it correct? Is it a temporary phenomenon that reflects passing international and domestic circumstances, or has it deep roots in fundamental changes in society and politics? Does it portend fundamental changes in public expectations of the armed forces? Are there any practical consequences of this perceived aversion to casualties?

We are looking at more than simply a concern to limit the loss of life in war. Napoleon and the generals of World War I apart, most states most of the time have sought to keep their own casualties down—not to zero, but to no more than good strategy and tactics demand. From ancient Greece and Rome to the absolute monarchies of the 18th century, soldiers have been costly to raise, train, maintain and replace, an understanding that revived after World War I. The wars of Napoleon and of the Western Front were exceptions, not the norm.

Nor is concern to save the lives of soldiers through increased medical support evidence of the phenomenon in question. In the past 150 years the revolution in medical affairs (yet another RMA) has been applied to war with great effect, reducing deaths from battle and disease quite remarkably. Antibiotics and the rapid evacuation of casualties for treatment have saved countless lives. Modern armies devote major resources to this purpose as a matter of course.

Nor, finally, is concern over casualties to be deduced from the desire to substitute technology and machines for men. Advanced states have always sought to exploit any comparative advantage in explosives, armour, artillery, warships, aircraft, bombs, missiles and so on. As the cost of recruiting, training and retaining manpower has risen, particularly since the decline in conscription in the West, there has been a natural tendency to substitute machines for people. Technology has also worked to put soldiers at ever-increasing distances from the direct line of fire.

For present purposes, aversion to casualties is defined as an increased reluctance on the part of democratic states to tolerate casualties by hostile action in military operations, a reluctance that is liable to significantly affect policy decisions. Before examining the connection between casualties and public attitudes, however, several notes of caution are necessary.

First, this chapter relies on public opinion polls, which must always be treated with care.[10] Much depends on the nature of the questions asked and their clarity or ambiguity; on the reliability of the sample; on whether the identical question was asked in the same way over a period of time; on whether questions were about an existing operation or a prospective or hypothetical one. Rarely does one find a complete and consistent record of opinion in a given conflict. Nonetheless, alternative methods of assessing opinion— insight, hunches, seats of pants, gut feelings and so on—seem even more unreliable.

Second, most of the examples in this chapter relate to the USA. This is in part because extensive poll data are available, in part because American reaction to casualties is more important than that of any other single country. What is true of the USA, however, seems likely to hold good in some measure for all established democracies.

Third, in any conflict there will almost certainly be a surge of public support at the outset—the so-called 'rally-round-the-flag' effect as people instinctively get behind the nation's soldiers. Unless the war is over quickly, however, public opinion has room only to shift against the war. It may well do this even if casualties are non-existent or minimal, as people become impatient with deadlock or any delay in achieving a result. At the same time, casualties can only grow over time or at best remain static. There is also debate on the best method of reporting casualty levels in relation to public opinion—the cumulative total, the logarithm of the total, or the marginal change in casualties each month or given period of time.[11]

Fourth, if casualties rise and public support falls for a given conflict, this demonstrates a correlation—not necessarily a cause and effect. Many commentators simply assume that 'the casualty factor' accounts for every decision not to use force, not to continue an operation or not to escalate military action. Yet failure to achieve objectives, the withdrawal of allies, the conclusion that the conflict is not an important interest—and many other factors—could also account for the waning of support.

Finally, those expressing opposition to government policy in a given conflict may want not withdrawal but escalation. In the Vietnam War some Americans at least responded to rising casualties with a demand for escalation of hostilities in order to defeat the North.[12] The fact that lives have already been lost in a conflict may indeed stimulate support for more decisive military action, even at the risk of greater casualties.

The rest of this chapter will examine the four questions outlined at the start of this section:

- What in practice has been the impact of public opinion concerning casualties on US policy?
- Are there deep-seated changes in society that might cause major shifts in public opinion for the long term?
- Are public expectations of the armed forces and the nature of military service changing?
- What are the consequences of the perception, if not the reality, of an aversion to casualties?

THE IMPACT OF PUBLIC OPINION ON US POLICY

In all Western democracies in World War II public support barely wavered, despite rising casualties and military setbacks. The war was a direct threat to the homeland and to each nation's way of life. The wars in Korea and Vietnam also began as major commitments for the USA, where clear national interests were perceived to be at stake in the fight against communism. Public opinion was initially prepared to accept casualties, but in both wars support fell as the number of deaths rose. According to one calculation, a rise in US casualties by a factor of 10 paralleled a fall in public support of approximately 15 per cent.[13]

The emergence of opposition to the war in Korea, however, can most plausibly be ascribed to the military stalemate that set in during 1951. In the case of Vietnam, domestic opposition can be attributed to the failure to achieve stated goals and to growing doubts about the cause for which US forces were fighting. The problem for President Johnson was not so much the level of casualties and opposition to the war as his conclusion that neither current efforts nor even a major escalation of hostilities would achieve a result.

The painful experience in Vietnam encouraged the USA to develop 'doctrines' concerning the use of force in an attempt to establish clear criteria for when troops should be committed overseas. These, it was hoped, would ensure public support if lives had to be put at risk. In 1984 Secretary of Defense Caspar Weinberger enunciated guidelines that were 'essentially rules for avoiding another Vietnam'.[14] As well as reasonable assurance of domestic support, they included defence of vital interests, clear political and military objectives, and an intention to win once engaged.

The Gulf War came close to fulfilling these criteria. Opposition before hostilities broke out focused on whether there were alternatives to war, such as economic sanctions, and whether the cause—punishing aggression, defending a friendly power or securing oil supplies—was a worthy one. Clearly, President Bush's purpose did not measure up to defeating the advance of communism but did win endorsement from Congress and a majority of the population. Polls that hypothetically suggested 1000 battle deaths showed about 50 per cent support for the planned action, while as many as 40 per cent said they would accept up to 10 000 casualties.[15]

Where Korea and Vietnam lasted for several years, the ground war in the Gulf lasted a mere 100 hours. If hostilities had continued for some months, however, opposition might have focused on casualties and support would probably have fallen away.[16] But the obvious conclusion needs qualification. A focus on casualties may simply reflect the best way for existing opponents of military action to get attention and press their case. Again, growing casualty lists might have led many Americans to press not for withdrawal but for greater military effort. Perhaps Saddam Hussein was fortunate that he did not turn the Gulf War into a killing ground for American soldiers as he had promised. One response might well have been escalation, even to the nuclear level, as the restraints present in Korea and Vietnam were lacking.[17]

While President Bush had moved away from formal doctrines on the use of force, the Chairman of the Joint Chiefs of Staff under President Clinton, General Colin Powell, returned to cautious and specific criteria for military operations, such as quick, decisive action and clear exit points. Public acceptability was clearly in mind as well as concern to avoid military entanglements that would drain resources and cost lives, as some of the new peace operations threatened to do. This was echoed in Clinton's Presidential Decision Directive no. 25 of May 1994 on US involvement in peacekeeping, which set as one condition 'domestic or Congressional support exists or could be marshalled for such a mission'.[18]

The USA's military interventions after the Cold War—notably Somalia, Haiti and the former Yugoslavia—seemed to provide stronger evidence for an aversion to casualties, whether in anticipation or after the event. Yet the picture is less clear on close examination. By the beginning of October 1993 the US public had already lost enthusiasm for continued participation in Somalia. Seven deaths had occurred in August and September, but other factors making for discontent were that the task of relieving the famine had been completed, social and political disorder did not seem America's business, and allies were not taking their fair share of the burden. Congress was already threatening to cut off funds by 15 November.[19]

The killing of 18 US personnel in Mogadishu on 3 October served as a catalyst. President Clinton set 31 March of the following year as the deadline for US withdrawal but at the same time sent 1700 fresh troops with 104 armoured vehicles: 'So now we face a choice. Do we leave when the job gets tough, or when the job is well done?'[20] Public opinion was certainly shocked by the loss of 18 soldiers. However, support for involvement did not wane simply because of this loss—it was already declining. In fact, polls showed that only 37–43 per cent wanted immediate withdrawal after the deaths, and many of these based their view on the belief that the original goal of famine relief had been achieved.[21]

Part of the US public, indeed, wanted not withdrawal but increased involvement, with certain polls indicating a majority in favour.[22] The purpose was to step up the military effort against the warlord Aideed—perhaps out of revenge, perhaps out of a desire to bring an outlaw to justice, perhaps simply 'to get the job done'. The strongest pressure to pull out, moreover, came less from the public than from Congress, which, reacting to the strong and vocal views of a minority, believed that the bulk of the population wanted out of Somalia and in turn put pressure on the White House. President Clinton's decision to increase the commitment while setting a date for withdrawal sought to keep all sides happy.

The US Administration's hesitation to commit forces to Haiti later in 1993, symbolised by the failure of the USS *Harlan County* to enter Port-au-Prince, appeared to confirm the importance of casualties in decision making. It was reported that President Clinton resolved to go into Haiti only in a 'permissive' environment (i.e. minimal risk of casualties).[23] The Department of Defense endorsed this policy, having all along opposed sending forces to restore to office the deposed President Aristide, whom it considered psycho-pathic.[24] Nor did the public have any great enthusiasm for an expedition to Haiti so soon after the 18 deaths in Somalia. Yet fear of casualties may not have been less decisive than the evident lack of consensus among decision makers. The USS *Harlan County* episode in fact caused a sense of humiliation in some quarters, prompting demands for an examination of invasion options.[25]

Eventually the USA did commit forces to Haiti, but it did not do so in the case of Rwanda. Here was a country of even less strategic interest than Haiti or Somalia that might take American lives and resources. Fear of casualties certainly played a part, but there were other reasons to avoid taking on responsibility to stop the genocide and bring law and order to a distant country. European governments also feared an open-ended commitment, but two democracies—Australia with 309 personnel and Canada with 128—did contribute to the UN operation.

US involvement in the former Yugoslavia was marked by several years' hesitation and national agonising in which fear of being bogged down in 'another Vietnam' was often voiced, but again other factors were at work. There was a belief that former Yugoslavia was essentially a European problem and that European states should take the main burden. There was justified concern that precipitate military action might aggravate hostilities. When commitments were made, specification of clear end-points served the purpose of putting the warring parties and US allies on notice that America's presence could not be counted on indefinitely.

That said, the so-called 'Somalia syndrome' was clearly at work in the choice of an air offensive against the Bosnian Serbs in 1995. Later, extension of the US presence in Bosnia was justified by pointing out 'no casualties so far', while unwillingness to arrest war criminals was due in part to fear of casualties.[26] The choice of aerial bombardment to bring President Milošević to heel over Kosovo in 1999 certainly appealed as a 'minimum casualty' option but it is possible that a ground invasion would have eventuated, if only to avoid loss of credibility. Britain certainly appeared anxious to prepare for this possibility.

INTERPRETING THE RECORD

The record of US policy bears a more complex interpretation than simple aversion to casualties. Two critical factors seem to shape the extent to which Americans will tolerate casualties: the cause for which the operation is conducted, and the strategic and tactical calculus. These interact with two further factors: the degree of consensus at home and the role of the media. All may be influenced by the prospect or reality of casualties, but the extent of that influence will depend on many circumstances.

As far as causes are concerned, an approximate hierarchy can be suggested with the more influential ones first:

1 defence of the homeland, as in World War II;
2 defence of allies and vital national interests, as in Korea and Vietnam;
3 punishing aggression and promoting international stability, as in the Gulf;
4 preventing ethnic cleansing or genocide, as in the Balkans; and
5 humanitarian assistance where law and order has broken down, as in Somalia or Rwanda.[27]

The cause appears to have greater merit if it is endorsed by the United Nations or an international body.[28] Geographical proximity also helps, as in the case of Haiti in 1993 and Panama in 1989 (which was presented as a source of drug supplies to the USA). The fact that Kosovo was located in the midst of NATO countries similarly added legitimacy to NATO action in 1999.

The calculus for any military action takes in numerous inter-connected factors, including the likelihood of success, the expected duration, the availability of non-military alternatives, support from allies, the presence of US citizens or prisoners of war, the presence of a villain on the scene (who provides a tempting target) and so on. The expected or actual level of casualties is one among all of these. Its salience will reflect the strength of other factors making for action or inaction, continuation or withdrawal.

Both the cause and the calculus will enter into decision making and help shape public opinion. Broadly speaking, the more worthy the cause, the less the calculus will matter, including casualties; and the better the calculus—low-risk, short, successful action—the less essential it is to have a major cause to fight for. As these factors feed into decision making, they will be mediated by two further elements: the existence of consensus or disagreement among decision makers, and the role of the media. The more critical of these is the former.

Tolerance of high levels of casualties is associated strongly with consensus among decision makers and among the major political parties. This will normally be sustained in wars for defence of the homeland, but may begin to show cracks in other conflicts where the cause is less urgent and where the strategic and tactical calculus becomes less favourable. Once divisions begin to appear, then casualties can serve as a rallying cry for the opposition. This is not to deny a genuine concern for casualties and a real impact on public opinion but to emphasise that opposition to beginning a conflict or staying in it can be based on a wide range of additional factors.

The media will shape popular thinking about war, but it is important not to overestimate its independent influence. First, the idea that the media can determine policy is one that the media themselves are only too ready to propagate, especially among political leaders. It is also contradicted by the theory that governments and elites can 'manufacture' consent through the media.[29] Moreover, where there is political contention among decision makers the media may be exploited by those seeking to push the government in a particular direction. For example, US intervention in Somalia was not the result of an autonomous 'CNN factor' but rather of pressure from powerful elements in Washington, who ensured that the famine received media attention in the first place.[30]

There is much evidence to suggest that for the most part the media reflect public opinion rather than create it. Subsequent analysis of attitudes during the Vietnam War points towards this conclusion, undermining the claim that the media 'lost the war'.[31] In relation to Somalia, analysis of polls after media coverage of the abusive treatment of the dead Americans suggests that few opinions were changed from one side to another—rather, existing attitudes were reinforced.[32]

Overall, the level of casualties is clearly a legitimate concern in political and military planning, but it is one among a wide range of factors. Given this, the role played by aversion to casualties is liable to be enhanced if: (i) the cause for which the country is committing troops is not critical; (ii) the strategic and tactical calculus is uncertain; (iii) there is a lack of consensus among decision makers; and (iv) casualties, for whatever reason, become a subject which the media uses to criticise existing policy. Public aversion to casualties can be as much a dependent effect as an independent cause.

LONG-TERM SOCIAL CHANGE

The argument that aversion to casualties influences the policy of democratic governments is based largely on the experience of the past 60 years. It is possible, however, that the greater salience of the casualty factor is due primarily to a historical and incidental change in the nature of the conflicts that the USA has been engaged in—from world war for national defence, through Korea and Vietnam fought for great strategic interests, to lesser operations after the Cold War such as Somalia and Haiti. Over time the vital interests of the USA have been less prominent, and in consequence concern over casualties has grown. From this perspective any attack on the US mainland or on vital interests would see a return to a high readiness to accept casualties.

This conclusion is not inconsistent with the argument that there are also fundamental developments in Western society that will increase the importance of the casualty factor in future. Demographic changes, growing concern for personal security, distrust of governments, the decline of great causes and ultimately the end of war have all been mooted. While some believe such trends are already having a strong impact, others see a growth in influence only over a long time scale.

The tyranny of demography

The best-known argument is that of Edward Luttwak, who links growing aversion to casualties to the lower birth rates and smaller families of postindustrial society:

> a certain tolerance for casualties was congruent with the
> demography of preindustrial and early industrial societies whereby
> families had many children and losing some to disease was

entirely normal. The loss of a youngster in combat, however tragic, was therefore fundamentally less unacceptable than for today's families, with their one, two, or at most three children. Each child is expected to survive into adulthood and embodies a great part of the family's emotional economy.[33]

Add to this the claim that the size of families will continue to decline in Western societies and the case for growing casualty aversion becomes stronger. One estimate, for example, suggests that in Italy in 2050 some 60 per cent of offspring will be from one-child families.[34]

The argument is plausible, but qualifications are in order. It is difficult to measure the presumed change in attitude with any confidence. The argument also needs to consider the type of war that is fought. Families do seem to take into account whether a son dies in defending the homeland or on peacekeeping in a remote part of the world. Patriotism may still be thicker than parenthood.

The Luttwak thesis also assumes that the attitudes of parents will be reflected in government policy. This is reasonable in democratic societies, but how long does it take for the phenomenon to become embedded in policy? After all, the fall in birth rates in Western countries has been under way for some decades. Perhaps, too, changes over time are more important than absolute levels. The US birth rate, as Moskos points out, is higher than that of the former Yugoslavia where casualties seem more acceptable.[35]

An interesting question is whether the same phenomenon affects non-democratic regimes in certain circumstances. Luttwak suggests that India, which has an advanced elite, became casualty-averse during its intervention in Sri Lanka in 1987, as parliamentarians demanded withdrawal before losses rose even higher.[36] The leaders of China may be particularly sensitive to the demographic consequences of losing significant numbers of single children. Insofar as this is the case, that country's one-child policy will be the greatest force for peace in Asia for many years.

A related demographic argument looks at the changing age profile of the population in the developed world. In the next 30 years the proportion of people aged over 65 will rise from 14 per cent to 25 per cent, and to nearly 30 per cent in some countries.[37] This will have several consequences. The proportion of working taxpayers to non-working pensioners will shift from about 3:1 to about 1.5:1. The birth rate will diminish as the child-bearing group becomes smaller in relation to the rest of the population, further skewing the age balance. The elderly—especially the 'old old' (over 85)— are likely to demand an ever-greater proportion of the nation's resources to provide medical and other services.[38]

Armed forces will face dire problems. Younger workers will be in great demand in the civilian economy in order to provide the wealth needed to support the elderly. Government budgets will be directed more and more towards medical attention and care for the old, which will put greater pressure on defence spending. The military will face growing competition for skilled and intelligent young recruits while having to defend its budget against the grey hordes. Both demographic and financial trends thus point towards a growing reluctance to commit forces to hostilities that might cause the loss of personnel.

Personal security, not national security

The concern that citizens will become too interested in their personal affairs and too little involved in the interests of the state goes back at least to the 18th century. Immanuel Kant welcomed recognition of the rights of citizens in a true republic, but foresaw that individual freedom could lead to neglect of duty to the community. In the 19th century Alexis de Tocqueville, writing about the USA, claimed that the growth of personal property would undermine the martial spirit among citizens and reduce their readiness to submit to military discipline.[39]

It may be only in the second half of the 20th century that the majority of citizens in wealthy, democratic countries have come to enjoy sufficiently high levels of personal wellbeing for this phenomenon to take widespread effect. Disease, malnutrition and lack of medical care have virtually disappeared as major causes of death, especially among the young. Life expectancy has risen significantly in the past 50 years, in part because of the absence of major wars. At the same time, the waning of religious belief in an afterlife places a greater premium on maximising a person's time in this life.[40]

People concerned to maximise their wealth and wellbeing see threats in crime, drugs and harm to the environment; and they expect high standards of medical attention, education and personal safety. One indication of this is that in many Western countries the number employed in private security firms now exceeds those in the military or the police. In short, greater awareness of the value of one's own life seems likely to focus efforts on maintaining personal wellbeing and to undermine willingness to sacrifice it for the state. There is too much to lose.

The movement to expand and protect human rights reinforces these attitudes. Most Western countries have legislation that guarantees the political, social and economic rights of individuals. Official imprimatur is thus given for citizens to defend those rights, not least the right to live. Few actions of government are more potentially damaging to an individual's rights than to send him or her off to war.

Democracy and distrust

Kant was also an early exponent of the theory that democratic states are inherently peace-loving. One reason for this is that, as the general public suffers most in war, popular opinion will restrain governments through a free opposition and a free media. Another is that the principle of equal citizenship in a democracy implies that no person's life is worth less than others. There is no underclass, no peasant class, and no proletariat that might be regarded as dispensable.

Of course, democracies have shown in the past that they can tolerate high levels of casualties. The point is that the principle of equality among citizens has been and is being steadily expanded in democratic societies, and that this may in the long term impinge on war. For example, the emerging idea that every citizen is entitled to life-saving medical care regardless of the cost reinforces the value placed on individual life by the community and suggests a greater reluctance to sacrifice the lives of citizens in war.

Another characteristic phenomenon in democracies is growing mistrust of governments. This has always been part of healthy democratic politics, but war itself has helped raise suspicion to new levels. The Vietnam War demonstrated to many Americans that their leaders could make fundamental mistakes at the cost of many thousands of lives. Robert McNamara's belated admission that as Secretary of Defense he prosecuted the war even though he knew it could not be won served only to exacerbate doubts about the judgement of national leaders.[41] Who, then, can be trusted with the lives of young citizens in time of war?

Public unease about government policies has been evident in other ways. The British government proved obstructive and unhelpful when parents of five servicemen killed by friendly fire in the Gulf War sought to discover exactly what had happened and who was responsible.[42] Similar official prevarication over the effects of the use of Agent Orange in Vietnam and the possibility of a 'Gulf War syndrome' due to chemical or other toxic agents only heightened public suspicion. The truth of these claims is not relevant here. What matters is the widespread belief that the lives of those in uniform were squandered or blighted for reasons not easily explained or justified.

The decline of great causes

Western history has been marked by great causes, for which men have been willing to lay down their lives: the propagation of religion; the creation and maintenance of empires; the glory of the nation; the defence of freedom against authoritarianism, fascism, Nazism and communism. Especially after 1789 the state was able to call on its citizens for great sacrifices. World War I was fought not only to defend national boundaries but to create a 'land fit for heroes to live in' and to put an end to war itself. World War II was fought to make the world safe for democracy and, in the words of the UN Charter, 'to save succeeding generations from the scourge of war'.

The Cold War provided one further great cause: to defend democracy from a barbaric and expansionist political ideology. Wars in Korea and Vietnam were adjudged worth the price, at least at the beginning. Only when the USA realised that Vietnam was more about nationalism (which it could do little to counter) than about the inevitable march of communism was the field abandoned.

The end of the Cold War meant the end of a feared enemy and of a great cause. Other people's conflicts, barbaric and uncivilised as many of them are, seem far less likely to inspire democratic societies to send large numbers of soldiers to fight or risk their lives. This calls for motives that are humanitarian rather than patriotic—for love of a neighbour (often rather distant) rather than hatred of an enemy. There are indications that Western societies (and to some extent governments) are coming to see the role of armies in this light.

The end of war?

Parallel with the greater interest in the humanitarian role of armed forces is the notion that war itself is coming to an end as a social institution. This is a development based first on a secular growth of opposition to war on principle. Individual pacifists and scepticism about war have existed as long as armies, but only in the 1880s did the first organised and broad-based opposition to war as a social institution appear. Moralists and Marxists, Quakers and progressives formed a motley coalition, arguing variously that war was un-civilised and barbaric, futile and counterproductive, or a conspiracy against the working class.[43]

More convincing among the population at large has been a second strand of thinking, namely that war is too costly in terms of blood and treasure. Enthusiasts for war prior to 1914 believed that a future war would be brief, noble and bearable. It was a myth all too easily shattered. The advent of weapons of mass destruction in 1945 reinforced the costliness of war. Not even the greatest of powers or the dumbest of dictators could contemplate nuclear war with equanimity, or even a war that risked escalation to the nuclear level. War in this form at least had become not wrong in principle but simply unthinkable.

Perhaps more convincing among governments has been the idea that war does not pay, that it has not lived up to expectations as an instrument of policy. Russell Weigley argues that the 'recalcitrant indecisiveness of war' was already evident by the early 1800s.[44] The wars of the 19th century that delivered profitable victories for their originators seemed to disprove this thesis. But the 20th century has seen few aggressors keep their gains—from World War I to the Gulf War. War, John Keegan argues, 'may well be ceasing to commend itself to human beings as a desirable or productive, let alone rational, means of reconciling their discontents'.[45]

Many long-term trends thus point in the same direction— a greater unwillingness by individuals to see lives sacrificed in war and a growing uncertainty about the value of war itself. Some of the trends are certainly reversible: attitudes to war may turn around in the event of aggressive behaviour by the enemies of democracy. Other trends such as demography and the concern for personal security seem more deeply embedded. The spread of democratic principles to new countries and the expansion of wealth in most nations serve only to reinforce the process.

PUBLIC EXPECTATIONS OF WAR AND ARMED FORCES

Long-term changes in society and politics may affect not only attitudes towards war and casualties in general, they may already be influencing specific public attitudes towards armed forces. Concern about the composition of armed forces, the protection of the rights of individual soldiers, and the expectation that military service will be a relatively safe occupation, may shape public and government attitudes towards the actual employment of the military as an instrument of policy. Ultimately, use of the military may seem too destructive—not of people and property but of the rights of those in uniform.

CONSCRIPTS, VOLUNTEERS AND RESERVISTS

Public thinking about armed forces and their use reflects in part how they are formed. The ability of the state to sustain heavy casualties in war originated in the system of conscription developed in France after 1789. Inefficient and unreliable levies were replaced by a system that employed national and local government administration to provide armies with a steady flow of recruits.

For the next two centuries Western governments came to rely heavily on compulsory service to fight major wars. In World War I conscription was necessary to replace the vast losses of young men in all Allied countries save Australia, which rejected it in two referendums. While the public suffered these losses with a heavy heart, few questioned their necessity or their nobility. Despite greater concern to spare lives in World War II, public support for compulsory military service rarely wavered.

Conscription for the defence of homeland and of vital national interests remained part of public life in most Western countries. However, not all conflicts justified the death of conscripts. The French abandoned Indochina 'once it became clear that they would need draftees to continue the war'.[46] It was the change in the system of getting people into uniform that signified the point where public opinion found the level of sacrifice unacceptable. In general, democratic publics have seemed more willing to tolerate casualties among volunteers and less willing where conscripts and especially reserves are concerned.

In Vietnam the USA was prepared to use conscripts, but drew the line at mobilising the reserves and the National Guard.[47] Public opinion eventually found the loss of draftees in Vietnam unacceptable; but it is likely that deaths among part-time volunteers would have caused even stronger and earlier public reaction. Subsequent policy in the USA left the all-volunteer force reliant on reservists for major engagements such as the Gulf War.[48] The thinking was that if reservists are to suffer casualties, governments will ensure that the cause is good and the operation successful. Like the USA, Australia employed conscripts in Vietnam and did not seriously consider calling up reserve forces. The first conscript death, however, caused considerable outcry, in part because of controversy over the war and conscription in general, and in part because of the lottery used to select the small proportion required for service.

At present, most Western countries have abandoned or are moving away from conscription, and sensitivity to the loss of lives among volunteers appears to be growing. The argument is now more often heard that society is obliged to spare the lives of those that have volunteered for military service simply because they have volunteered. Volunteering for military service is seen as a major sacrifice in itself: not too much more should be demanded.

Individuals in the military

Increasing attention to the rights and status of individuals in uniform stems in part from changes in the wider society, and in part and more directly from the need of the forces to recruit, motivate and retain people who expect more regard for their rights than their predecessors of a generation ago. The greater the public attention to service personnel as individuals, it can be argued, the more difficult it is to see them as casualties.

Western democratic societies have long been concerned to protect the rights and freedoms of citizens as civilians. In recent decades, however, these have been extended to the military in a wide range of ways: gender equity, equal opportunity, freedom from harassment, occupational health and safety, rights to sexual preference, due process in legal proceedings and so on. All serve to emphasise the rights of individuals in the armed forces. Much public and political criticism of the military, indeed, focuses on failure to protect the rights of individual service personnel.

Other evidence pointing in the same direction is the attention given to the fate of those missing in action after Vietnam, which remains an issue in the USA over 30 years later. After the Soviet withdrawal from Afghanistan, subsequent governments displayed an unprecedented concern for prisoners of war still in captivity and for the 311 soldiers still missing.[49] The effects of war on those who survive is receiving more attention now than ever before. Understanding of post-traumatic stress disorder has greatly increased, and governments have realised—often after vocal demands for compensation—that ex-soldiers may require years of treatment for all manner of ailments and conditions. The prolonged sagas of Agent Orange in Vietnam and the 'Gulf War syndrome' linked to toxic agents demonstrate that the harm suffered by individuals can secure continuing publicity and considerable public sympathy.

Perhaps the most telling fact in heightening public awareness of the individuality of soldiers is the recording of names of all those soldiers killed in wars. This 20th century phenomenon has become possible with efficient administrative procedures, but the motive behind it is the desire of communities to honour their dead as individuals. World War I led to a proliferation of monuments naming those that had been killed, a tradition continued through to the Vietnam War and after. During that war Australia adopted the practice of bringing home the bodies of those killed rather than burying them in overseas war graves as in the two world wars. Air transport and relatively low numbers made this possible, but the effect was to make the community more aware of the personal sacrifices involved in war.

Concern for individuals is manifest, too, in the protection accorded to those with a conscientious objection to participation in war. In the course of this century Western democracies have expanded the rights of objectors, moving from a religious basis to a more secular notion of conscience.[50] The principle is that citizens cannot be made to do something that is morally abhorrent to them. Australia has gone one step further in giving legal protection to those that have a conscientious objection to fighting in a particular war. The relevant statute applies to those conscripted on some future occasion, but the principle can be extended to those already in uniform.[51] There is no logical reason why a soldier that forms a genuine objection to a particular war should have his conscience overridden—if protection of individual rights is paramount.

The sociology of the military

The all-volunteer force now found in many Western societies has led to a growing public interest in the sociological composition of the armed forces. As membership becomes more diverse, more reasons exist for the community to be concerned about who joins the armed forces and who is likely to lose their life. The concern is not just how many casualties but who might be a casualty. Several grounds can be identified (see below).

Bias in the casualty list?

The steady lengthening of the casualty lists from Vietnam caused many in the USA to ask whether the burden was being unfairly borne by certain social groups, in particular those that were poor and black. While the wealthy could find ways of deferring service by prolonged university study or by leaving the country, the poor seemed more likely to find themselves in the military—and in the firing line, where their relative lack of skills often led them. Significantly, too, the poor included disproportionately more blacks.

Intense debate arose over whether black soldiers were being killed in Vietnam more than proportionately to their numbers in the army and in the community. The data show that, despite some extreme claims, blacks made up 12.1 per cent of all US personnel killed in South-East Asia—a figure comparable to the proportion in the community at large and slightly lower than the proportion in the army at the end of the war.[52] What is important here is that the issue stimulated public concern for the identity of casualties, a concern that can only enhance any existing aversion to casualties.

Women in body bags?

Many all-volunteer forces are increasingly dependent on the recruitment of women. Inevitably, controversy has arisen over whether women should be permitted to take part in direct combat and whether the public would accept the idea of women being killed, wounded or taken prisoner. The deliberate exposure of women to such fates seems to offend against the traditional and deep-seated image of war in all societies. No public, it was argued, would tolerate women 'coming home in body bags'.

The Gulf War seemed to belie this concern. Eleven women died, five through hostile action, and several were taken prisoner. The media paid great attention to these cases, but the US public did not turn against the idea of women in combat. Overall, the Gulf War served to dispel the fears of nervous policy makers and encouraged the public to see women as a normal part of the professional military.[53] Both the media and political leaders may have misread public attitudes because of their own image of the male warrior,[54] or because they assumed that image was still deeply entrenched in American society.

How the public would react to female deaths numbering, say, in the hundreds is yet to be tested. The Gulf War was exceptionally low in Western casualties. Many of those who oppose allowing women into ground combat claim that public reaction to female deaths on a larger scale would cause demands for their withdrawal from the firing line and perhaps the nation's withdrawal from the war. Only experience will tell whether the reaction will be different in kind from the response to the 240 men killed in the Lebanon in 1983 or the 18 killed in Somalia in 1993.

Should married couples and parents be sent to war?
Conscription ensures that a military force is predominantly young and unmarried. An all-volunteer force generally contains a higher proportion of married personnel, including couples that are both in uniform. There are indications that the US public is more sensitive to the loss of married personnel than single, especially when both partners are at risk and when children are involved. During the Gulf War, *People* magazine produced an issue featuring a picture of an Air Force woman holding an 11-month-old baby, with the caption 'Mom Goes to War'.[55]

Sending single parents to a war zone is even more sensitive, especially when linked to the much-publicised failure of some to provide adequately for dependent children during their absence in the Gulf. Over 26 000 single parents deployed with US forces, including 4000 women. Of the total 375 combat and non-combat deaths associated with *Desert Shield/Desert Storm*, six were single parents (five men and one woman).[56]

There was some political and public expectation that US forces should not send single parents or parents with young children into a war zone. After considerable debate following the war the three services adopted common rules, which provided for a four-month exemption from combat assignments for new mothers, single parents and one member of a military couple.[57] From the military perspective this complicates deployments and postings, but it seems to be a price that has to be paid, given public and political concern over sending parents and married couples to war.

Where are the elite?
Charles Moskos attributes the greater reluctance to accept casualties in the all-volunteer force in part to the fact that the children of the elite are no longer to be found in the ranks of the armed forces.[58] Conscription, by contrast, ensured that many, if not all, of the more privileged youth were compelled to serve. Well-born young men played their part in the two world wars and in Korea, but many successfully avoided or evaded the draft for Vietnam,

thereby helping to undermine popular support for compulsory service in Vietnam.

Only when children of the elite accept their fair share of the sacrifice, Moskos argues, will the community be more willing to accept casualties in wars where great national interests are not at stake. While this helps explain growing concern for casualties, it does not necessarily follow that compelling the children of the elite to put on uniform and share the risks will solve the problem. As the draft for Vietnam demonstrated, privileged youth can usually escape conscription one way or another.

The safe military

For the past 20 years, service in most Western armed forces has been a relatively safe occupation. One of the least likely ways of dying has been through hostile action, as Table 4.1, based on all deaths in the US military from 1980 to 1997, illustrates. The figures for hostile deaths include 256 in Lebanon during the period 1982–84; 18 in Grenada; 23 in Panama; and 148 in the Gulf War. Over 18 years the average is 31 per annum, and this has been falling since 1991. Of course, there is no guarantee that this trend will continue, but the longer it lasts the more it will reinforce the perception that military service is safe and that personnel will not be sent in harm's way. The British Army found that it had to make clear to new recruits that they might expect to be shot at and they might be required to kill people.

Accidents, fatal illness, suicide and homicide are part of civilian as much as military life; this can be understood and accepted. However, deliberately exposing soldiers to danger seems another matter. One indication of this is the anger on the part of some parents whose children, having volunteered for service in peacetime with no real expectation of having to fight, are put in danger. Luttwak uses the Italian word *mammismo* or 'motherism' to describe this phenomenon.[59] The effect is no doubt magnified if soldiers are able to phone their parents from the war zone.[60]

At the same time several other occupations outside the military are becoming more dangerous. Between 1989 and 1998 some 472 journalists were killed in pursuit of their profession.[61] In the first half of 1998, more civilian employees of the United Nations were killed in the course of duty than uniformed peacekeepers.[62] Workers for non-government organisations are also frequent casualties of political violence in civil wars. Such deaths do not seem to cause

Table 4.1 Worldwide US active-duty military deaths (1980–September 1997)[1]

Total deaths	31 331
Accident	18 643 (60%)
Illness	5 680 (18%)
Self-inflicted	4 256 (14%)
Homicide	1 738 (6%)
Hostile deaths	560 (2%)
Pending/undetermined	454

1   Department of Defense, Washington Headquarters Services, Directorate for Information, Operations and Reports, Table 1 (microfiche). Deaths in the Australian Regular Army 1977–91 (total 597) were in roughly the same proportions, though with more illness (25%), less suicide (11%) and homicide (1%), and no hostile deaths. See R.J. Thompson, P.G. Warfe & R.J. Lipnick, 'Causes of mortality in Australian regular army personnel, 1997–1991', *Australian Defence Force Journal*, November/December 1997, no. 127.

widespread public concern, perhaps because most volunteer to go in harm's way. While soldiers volunteer for military service, they generally do not choose whether or not to attend a war. This decision is the direct and immediate responsibility of the government and the community.

Safe, clean wars

Public ideas of war are difficult to pin down, but two images seem to dominate. Both militate against acceptance of casualties. First, there is high-technology war, epitomised in the Gulf War and in the Kosovo campaign of 1999. Precision-guided missiles, viewed by millions, travel down streets, aiming for a selected building, even a specific window. Military targets are identified and destroyed from afar. If aircraft are shot down, their crews are rescued by helicopter. Even the ground war in the Gulf produced remarkably few deaths among Western forces.

Another measure of the low casualties is the study commissioned by *Time* magazine that compared total deaths during the Gulf War with total deaths normally occurring over the same period in peacetime. It found that 40 per cent more soldiers would have died if the war had not taken place.[63] If this is the public expectation of war, it has set a very high—and almost certainly misleading—standard. Only wars that are a similar mismatch of capabilities are likely to be as safe for the winner.

This first image of war contrasts strongly with the second image: that of nasty, dirty wars, employing low-technology weapons in brutal and indiscriminate fashion, with civilians suffering and dying in great numbers. Local hatreds and animosities seem out of control, while political direction is lacking or, if present, pernicious. Such barbaric conflicts are seen as the dominant form of war in the present era, not only by the public but by academics and commentators.[64]

Whether soldiers are sent in to take sides or to mediate between warring factions, the rationale for commitment is easily questioned. The conflict does not represent a direct threat to the nation. A clear end-point is rarely visible or reached in practice. Deaths seem likely to go on indefinitely. Soldiers are killed by unworthy opponents using cowardly tactics. While soldiers require no less courage to deal with this kind of threat, it is harder for the community to glorify and accept their deaths. Being shot in the back or blown up by a landmine in a remote conflict does not translate easily into heroically giving up one's life for one's country.

THE CONSEQUENCES OF CASUALTY AVERSION

The hypothesis that casualties are increasingly unacceptable to Western democracies, and to the USA in particular, has a certain life of its own. True or false, it can affect the behaviour of three important groups: enemies, political leaders and the armed forces.

Enemies

If opponents of the USA or other democracies believe that public opinion at home is unable to tolerate casualties, certain strategic and tactical consequences may follow. For example, according to one senior military defector from North Korea, the government in Pyongyang believes that the loss of 20 000 troops in the initial phase of an attack from the North would suffice to make the USA abandon the South.[65]

At the tactical level an adversary has a similar incentive to cause US casualties. Luttwak argues with respect to Bosnia: 'No adversary, should any emerge, need fear that attacking US troops would only provoke the arrival of more US troops'.[66] If it is not possible to cause a large number of deaths—as was done in the blowing up of a barracks in Beirut in 1983—then vicious treatment of prisoners or abuse of dead soldiers might suffice. Tucker suggests that an enemy could defeat the USA 'by broadcasting to us live the mutilation of American prisoners of war, who of course in the future will include women, and then returning them mutilated and disfigured to us as a derisive goodwill measure'.[67]

A similar tactic would be to target US soldiers in a multilateral force. In Australia concern was voiced when Foreign Minister Alexander Downer expressed doubt about participation in a peacekeeping force in East Timor and referred to the undesirability of Australians coming back in body bags. Such an attitude invites an enemy to target the most politically vulnerable participants in a multinational force.[68]

Political leaders

Leaders of democracies are influenced by public opinion, or at least by what they believe to be popular sentiment over casualties. This is no new phenomenon. During the US election campaign of 1864 President Lincoln was concerned that high casualties 'in the absence of any clear gain on the battlefield, might lead a weary electorate to abandon him and the cause'.[69] Several problems for political leaders can be identified.

First, there is a danger that political leaders will see winning the support of public opinion as *the* critical test for undertaking military operations. In the Falklands War the British Government set great store on winning and retaining public support.[70] In the Gulf War President Bush was anxious to secure a resolution from Congress backing military action. To a point this is healthy in a democracy. The problem arises when courting popularity becomes the dominant concern of national leaders, rather than achieving the purposes of the war. This anxiety is fuelled by a readiness to over-rate the impact of the media, which are believed capable of whipping up popular reaction against a conflict that is producing casualties.

A related failing of leaders is their distrust of the public, a fear that they will not understand the reasons for the loss of lives in a particular conflict. In fact, democracies have shown much good sense, being generally willing to accept casualties where worthwhile issues are at stake and there is a reasonable calculus for action. The British public, for example, has not reacted against the numerous deaths suffered by its army in Northern Ireland since 1969. Democracies have also accepted a certain loss of life in peacekeeping operations—such as France in the former Yugoslavia. It is when governments fail to explain or cannot explain the merits of an operation that the populace is liable to become critical of casualties.

A final political failing in the face of the presumed intolerance for casualties is the tendency to demonise opponents in order to secure public support. The USA painted Noriega as a drug lord, Saddam Hussein as a new Hitler, Aideed as a primitive warlord, the rulers of Haiti as brutal dictators, and Milošević as the villain of the piece. Use of the armed forces is presented to the public as a hunt for the guilty individuals. If the quest fails, the public begins to wonder whether loss of life is worthwhile simply to bring a criminal or two to book.

Armed forces

What does the casualty factor mean for armed forces themselves? Three trends are evident: a focus on high-technology weapons; adoption of tactics and strategies that set minimal casualties as a high priority; and a shift in the values and ethos of military professionals.

Weapons
The RMA has already produced an expectation that weapons will become increasingly accurate and discriminating, even if fired from the other side of the world. The USA's cruise missile attacks on terrorist targets in Afghanistan and Sudan in 1998 reinforced this line of thinking, as did the precision bombing—more or less—of Yugoslavia in 1999. If the future of war is seen as operations from distant command bunkers or ships and planes standing off a target, this will serve to reinforce the expectation of minimal casualties. The focus will be on weapons that do not require many troops, if any at all, to go into the battlefield. 'Bomber pilots', the saying goes, 'do not have mothers'.

In the extreme, some believe, the USA could 'bring the troops home' and manage the world from its continental redoubt.[71] The pressure is for armed forces to acquire weapons that promise a bloodless war, in the first instance on one's own side but also in terms of less killing of enemy soldiers and civilians. The idea of 'surgical strikes' against a nation's 'infrastructure' is very appealing—'Bomb now, die later'. The public and political leaders may expect weapons that destroy property and systems, and neutralise through non-lethal technologies rather than kill people.[72] The RMA and aversion to casualties are close allies; each can be used to support the other.

Tactics and strategy
The principal concern for armed forces is that too high an emphasis on minimising casualties will adversely affect tactics and strategy. If taking casualties is seen as a black mark, commanders may come to focus unduly on force preservation. Means may be chosen more for their low level of risk than for effectiveness—which may require overwhelming force. The desire to avoid casualties is also likely to lead to a defensive mentality, an aversion to risk taking, the loss of initiative and the disappearance of creativity.[73] The result may well be greater casualties and a reduced prospect of victory.

Luttwak suggests—ironically, one presumes—the adoption of a 'casualty exposure index', which assesses the likelihood of casualties for each element in the armed forces: infantry presents the highest risk, armoured brigades rather less, and so on.[74] The long-range cruise missile is the weapon of choice, although monetary costs may be prohibitive. A precise calculus is needed to allow for the fact that too many casualties may cause an operation to be abandoned even if those forces with a high casualty exposure index are most suitable in military terms. This is the concern for casualties *reductio ad absurdum*.

The military ethos
Military leaders are expected to be solicitous of the lives and wellbeing of their soldiers. Too great a focus on avoiding casualties, however, is liable to lead not only to poor tactical and strategic decisions but to a decline in the ethos of commitment and sacrifice—qualities essential to any armed force.[75] If soldiering loses the ultimate readiness to sacrifice one's own life or to send others in harm's way, it loses its unique characteristic as a profession.

Concern to protect one's own forces may contribute to a decline in morale. A revealing episode was the withdrawal of Australian Army mine clearance trainers from Pakistan. The then Minister for Defence stated that one reason for the decision was the risk to Australian lives, a rationale that was vigorously criticised by the soldiers concerned. On peacekeeping operations, invidious comparisons may be drawn by other contingents. In Macedonia, for example, the preference of US forces for low-risk roles led to criticism from Scandinavian forces.[76]

The challenge for officers is great, especially when minimising casualties is part of war-fighting doctrine, as is the case in the US Army.[77] They must balance concern for soldiers' lives with achieving the aims of the mission. They must also retain the respect of the public at large, for it is to them that the lives of the young generation are entrusted. Failure in this regard only enhances public reluctance to accept casualties.

Another trend that may impinge on the military ethos in this context is creeping commercialisation. The growing use of civilian contractors and their advance ever closer to the front line may have the effect of changing attitudes towards acceptance of casualties in the armed forces as a whole. If special protection has to be provided for civilian contractors, the question arises why the same should not be done for soldiers. On the other hand, the public may be more tolerant of the deaths of civilians that have chosen to take part in war simply for profit.

The presence of the media also poses difficult problems. The family of a soldier killed in a distant country may well learn of the event from the media before national authorities can inform them. This does not promote acceptance of casualties. Soldiers on operations may also know that casualties are being reported back home and may become concerned about the effect of this reportage on family and friends. Again, this situation will enhance rather than diminish concern over casualties.

PROSPECTS

The perceived reluctance to accept casualties is linked to two other contemporary developments in the use of force by the state. One is the growing interest in the use of mercenary forces as a substitute for a nation's own volunteers. Luttwak, for example, suggests the Gurkha model (recruiting groups of foreign nationals) or the Foreign Legion model (recruiting individuals).[78] Either would be suitable for long-term service. Alternatively, governments may simply purchase mercenary forces for a particular operation on the open market. The effect of such policies is to minimise or eliminate the casualties incurred by one's own nationals. It may not be coincidence that the greatest use of mercenaries since 1945 appears to have been by democratic governments.

Another development related to the avoidance of casualties is the option of assassinating enemy leaders, a course of action that promises to achieve results with minimal risk to the lives of soldiers. Assassination also accords with the demonisation of adversaries noted earlier and, in that leaders are seen to be the cause of many problems, with current efforts to bring major war criminals before international tribunals. Democracies, however, have difficulty in openly espousing assassination as a policy, and in the case of the USA an Executive Order formally prohibits direct and indirect participation in such action.[79] Several military actions—from the bombing of Libya in 1986 to the attacks on Baghdad in the 1990s—might nonetheless have produced this result 'incidentally'. Actions of this kind may increasingly appeal to democracies as an instrument of policy that does not put soldiers at risk.

The use of mercenaries and of assassination represents a radical departure from the traditional ways in which states have used armed forces against one another. It may be, however, that governments have become too focused on preparing to conduct 'Napoleonic' warfare (i.e. with high casualties justified by a vital cause).[80] Such an approach does not easily carry over to low-intensity warfare, where the loss of soldiers is less easily justified. Over time strategic thinking, particularly in the USA, may focus more on small wars with low—but not zero—casualties. Countries such as France and Britain, by contrast, seem rather more ready to tolerate casualties as a result of their long engagement in colonial wars.

Armed forces have grown more concerned about soldiers' lives than ever before but not so far to the point of undermining their own ethos and *raison d'être*. Political leaders in democratic countries are still able to commit forces to conflicts but now face a greater degree of accountability for the lives of soldiers. The public justifiably wants to know that operations are for an important purpose and that they are competently conducted. This naturally becomes more difficult if decision makers are divided or uncertain, and if the media are active in voicing doubts.

Yet there are also deeper social and political trends that may steadily increase concern over casualties and alter how democratic societies perceive the armed forces and war. If these trends take hold, aversion to casualties may come to be so strong that it will radically change war itself.

5 | Post-traumatic stress and the Australian Defence Force: lessons from peace operations in Rwanda and Lebanon

Peter Warfe

THE AIM OF this chapter is to examine the stress to which peace-keepers may be exposed in contemporary operations. It argues that peacekeeping may be much more stressful than commonly believed. There is a notion held by many in society that peace-keeping is some sort of paid vacation in an exotic location. This is a misguided and erroneous assumption. At best, peace operations provide the most realistic experience of near-combat situations; at worst, they remind us clearly that peacekeepers and combatants may be exposed to extreme levels of shock, horror, fear, disgust and other psychological and physical stresses.

To illustrate the nature of peace operations, this chapter analyses two peacekeeping case studies: the first is the ADF's experience in Rwanda in 1995, and the second the Norwegian experience in Lebanon between 1978 and 1991. The chapter goes on to outline several of the initiatives the Australian Defence Force (ADF) has taken to try to prevent the incidence of both acute stress and post-traumatic stress disorder (PTSD) among service personnel. It concludes by suggesting that the ADF would be well served by a centralised agency designed to coordinate all research and analysis of acute stress and PTSD.

DEFINING OPERATIONAL STRESS: THE INTERFACE BETWEEN TECHNOLOGY AND THE HUMAN FACTOR IN WARFARE

In recent years, much defence debate on the future of armed conflict has focused on technology-driven solutions and has tended to gloss over the human implications of warfare. The growth of military technology, particularly in the fields of electro-optics, digitisation, integration and miniaturisation has raised the possibility of the soldier becoming a far more capable force in the field. There is a growing interface between battlefield operating systems and the soldier. Health issues that had previously been given a low priority, notably both physiological and psychological stress, have gained increased prominence.

When the Australian Army introduced Project Wundurra—now called Land 125—a few years ago, the Defence Health Service Branch was comforted by the statement that the project aimed 'to avoid technology for technology's sake'. Because of this focus, Defence Health has been able to introduce physiological, human factors and behavioural inputs to assist the project. Land 125 has continued to concentrate on maintaining performance levels under stress in order to enhance a soldier's survivability, sustainability and performance.

The entire history of warfare indicates the central importance of the soldier and the key roles played by preventive medicine and casualty prevention. The ultimate goal of preventive medicine is to ensure that commanders have sufficient personnel resources to accomplish their various missions. However, we must now look beyond factors such as physical disease and injury, because our forces will also be subjected to intense psychological stress caused by overwhelming activity levels on a 24-hour battlefield.

What is meant by the word stress? Stress is defined here as: the physiological and psychological reaction that occurs when individuals perceive an imbalance between the level of demand placed on them and their capability to meet that demand. Individuals experience stress as an internal, personal response that cannot be measured by onlookers. Accordingly, subjective comparisons of levels of stress suffered in peacekeeping operations as opposed to those experienced during war are largely invalid. Operational stress, or acute stress disorder, encompasses an array of effects caused by the strain of operations; the term refers to a usually temporary psychological upset, which causes a marked reduction in an individual's ability to function effectively. PTSD, on the other hand, is caused by a psychologically traumatic event that is generally outside the range of normal human experience, resulting in symptoms of acute mental and physical distress. Symptoms include intrusive flashbacks, emotional dissociation, psychological numbing, arousal of fear, depression and substance abuse. While there is a difference between peacekeeping and warfare, the nature and internal contradictions of peacekeeping underlie many recent cases of acute and post-traumatic stress reaction.

THE NATURE OF PEACEKEEPING

In principle, peacekeeping is different from peace intervention. Yet in practice the distinction between protecting and enforcing peace is often ambiguous and a matter of continuous interpretation and political controversy. Peacekeeping operations are designed to prevent conflicts from escalating to the level of major threats to international peace and security. They also aim to reduce or prevent intolerable human suffering. Paradoxically, the presence of a peacekeeping force often prolongs the conflict by freezing the status quo. Conventional wisdom states that peacekeeping must be based on consensus; the interested parties must prefer the presence of a peacekeeping force with its attendant constraints on freedom of action to unencumbered violence.

Clearly, the effectiveness of a given peacekeeping mission depends critically on its reputation for impartiality. This kind of reputation is shaped by the actual behaviour of the peacekeeping forces, also by their identity. Paradoxically, the effectiveness of a peacekeeping operation is often inversely related to the amount of force it employs. Individual soldiers are required to use restraint in the application of force; they are required to adhere to stringent, sometimes complex rules of engagement. They develop a fear of their own aggression, a situation that differs from soldiers in wartime operations. In addition, many peacekeepers hate being involved in the recovery of the dead—especially dead children or, alternatively, the sight of children being killed.

Successful peacekeeping depends largely on mutual respect and effective liaison with all of the contending parties in order to prevent a succession of incidents or misunderstandings from escalating into large-scale violence. The point to note is that peace-keeping is a tall order by any standard: it involves diplomacy and much military contradiction. Well may we say: 'blessed are the peacemakers'.

THE CASE OF RWANDA: THE KIBEHO MASSACRE

An escalating guerilla war in the central African republic of Rwanda culminated in genocide in April 1994. An estimated half to one million civilians were massacred and two million refugees fled the country, leaving about one million internally displaced people residing in internally displaced persons' (IDP) camps. Following the genocide of April, there was a three-month civil war that was won by the Rwanda Patriotic Front (RPF) in July 1994. In August, the United Nations passed a resolution to increase the world body's assistance mission in Rwanda to a 5500-strong peace-keeping force, called United Nations Assistance Mission in Rwanda (UNAMIR).

In the same month, the Australian Defence Force Medical Contingent was deployed to provide health care to UNAMIR. The ADF contingent's secondary role was to provide humanitarian support from within its spare capacity. The ADF contingent numbered 300 personnel, comprising both inpatient and outpatient medical and surgical services. A logistics company and an infantry company for self-protection supported the medical personnel. Australia maintained this level of support to UNAMIR for a year, rotating two contingents for a six-month tour of duty.

At Kibeho, the victorious Rwanda Patriotic Army (RPA) believed that many perpetrators of the April genocide along with members of the former Rwandan government's military forces had taken refuge in the IDP camps. For the RPA, the refugee camps represented sanctuaries for criminals that deserved the most severe punishment and constituted a potential military threat. Kibeho is an old Catholic mission station built along a ridge line. In the north, at the highest point of the ridge line, was a church. South of the mission station there was a Zambian Army company headquarters and platoon position. Still further south was an old mission complex that made up a compound.

The period 7–14 April 1995—the anniversary week of the 1994 genocide—was declared a national week of mourning by the RPA government. There were anti-UN demonstrations in the capital Kigali. Fear of revenge and retaliation by the RPA led to a dramatic swell in the population of the IDP camps as people sought safety. At 0300 hours on Tuesday 18 April 1995, two battalions of RPA troops surrounded Kibeho camp, creating a cordon. RPA forces fired shots in the air to intimidate IDPs and to herd them into the camp; one woman was shot in the hip and 10 people, mostly children, were trampled to death in the stampede of fear that followed.

IDPs running into protective concertina barbed wire suffered many injuries. Given that the Zambian Army company was the only UN force immediately available, the situation was highly unstable. Staff officers at headquarters UNAMIR worked frantically to provide reinforcements for, and medical support to, the Zambians. The headquarters also attempted to brigade all available UN transport assets to assist in the removal of the IDPs. On the following morning, Wednesday 17 April, a 32-person Australian medical team arrived at Kibeho at about 0930 hours. The team comprised medical and evacuation sections; there was an operations command post, while organic security was provided by two infantry sections. At 1000 hours the Deputy Force Commander, the Colonel (Operations) and I visited Kibeho to try to assess and defuse the situation on the ground. We found that thousands of people were packed along the ridge line, in an area about a kilometre and a half long and some 200–300 metres wide. We also observed that the bodies of the 10 IDPs killed in the panic of the previous day were still lying out in the open exposed to the elements. By Thursday 20 April, conditions inside Kibeho were deteriorating rapidly; the force commander visited the camp and held talks with UNAMIR troops and representatives of various UN agencies. At 1730 hours some IDPs attempted to snatch RPA weapons. The RPA forces reacted by opening fire, killing 20 people and wounding 60 more. There were also reports of IDPs fighting among themselves using machetes.

Saturday 20 April 1995 was the day of the Kibeho massacre. During the night of 19–20 April there had been a large number of killings. Many of the IDPs were in a poor physical condition, as no food had been distributed to the camp for five days. At midday, rain fell from an approaching thunderstorm and people in the camp rushed to find shelter, creating a panic. This sudden mass movement of people was interpreted by the RPA as an attempt to break their cordon. Immediately RPA forces opened fire into the crowd and continued firing for an hour, killing around 130 people. Throughout the entire day, the Australian medical team worked furiously, treating those victims they believed had a chance of survival. At 1600 hours, UN helicopters were finally granted permission to land at Kibeho with medical supplies and to airlift the wounded to hospital. At the same time, a 22-strong RPA platoon marched in formation down the road from the church, singing and chanting. The words they sang were, 'We killed the Hutu. Who will we kill next?'.

The RPA platoon stopped near the Zambian Army compound and cocked their weapons. Two Australian Army private soldiers in the bunker nearest the RPA thought that they were about to be overrun and killed. A lance corporal ordered his infantry section to fix bayonets; but the RPA ignored the Australians and began firing into the crowd, causing another breakout attempt by IDPs.

There was so much firing that the Australian medical team suspended work in the casualty-collecting post and sought cover in bunkers. When the crowd in the camp surged against the cordon, the RPA responded by opening fire with heavy machine guns and rocket-propelled grenades. A number of RPA soldiers moved among the fallen bodies, bayonetting to death or finishing off the wounded with gunshots. Many of the IDPs were rounded up and marched away as if under arrest; they were then summarily shot.

Australian infantrymen witnessed the slaughter. Although extremely frustrated at their inability to intervene to prevent the killings, they were determined to protect their medical personnel. At first light on Sunday 21 April, Australian medical personnel conducted a count of the injured and dead at Kibeho. Using pace counters, they found some 4000 dead and 650 wounded. The Australians also observed the RPA exhuming bodies in the camp and transporting them away. The aim was clearly to reduce the number of bodies to about 300, which was later the Rwandan government's official number of those killed. Australian medical personnel were denied access to the exhumation and reburial sites. However, we later learnt that the RPA had obtained transport to move the bodies to the nearby town of Butare, where they were burnt by a local contractor.

On Monday 24 April, approximately 1700 IDPs were reported as remaining inside the Kibeho compound. Throughout the day, Australian personnel ventured into the camp to remove the bodies of 42 more dead and to place them in a mass grave. This was a most unpleasant task, as dogs had gnawed some of the dead and rats had taken shelter inside some of the body cavities. Eventually, the remaining IDPs dispersed to their home communes where, sadly, once again many of them faced illegal jailing or death.

Many of our troops were angry and frustrated by the bloodshed they witnessed at Kibeho and their inability to intervene. As a consequence, the contingent put in place a comprehensive stress management program, which included debriefing by commanders, doctors, psychologists and the padre. In addition, shortly before our return to Australia, Army psychologists conducted group and individual debriefings. In Australia, all contingent personnel have been followed up by letter at the six- and 12-month marks.

I was particularly concerned about the feelings of anger and hatred that existed among Australian troops. There was always the possibility of individual soldiers exacting retribution from the RPA back in Kigali. If such incidents had occurred, they would undoubtedly have invoked a furious response from the RPA. I visited Kibeho after the mass killings and, while not condoning what the RPA had done, I tried to put the massacre into a perspective that fitted the overall tragedy of Rwanda as a war-torn society. I concentrated on the numbers of lives saved and on the fact that the presence of our troops had almost certainly prevented a full annihilation of about 50 000 people in the camp.

In the aftermath of Kibeho, we identified around a dozen troops that appeared to be having difficulties resolving the experiences to which they had been subjected. Since our return to Australia we have been following up contingent members, but it has been difficult to keep in touch with those that have separated from the ADF and those that have been referred elsewhere for ongoing care. This illustrates one of the major problems that confronts the ADF in such follow-up exercises—our inadequate system for capturing data. The incidence of PTSD in Rwanda veterans—about 20 to date—is consistent with PTSD rates in emergency service personnel, but it is lower than among survivors of civilian disasters. An important point to note, however, is that training and selection does confer some level of protection from acute stress reactions.

More than half of the Australian contingent served at Kibeho during the savage month of April 1995. The contingent's planning, presence, military discipline and compassion undoubtedly saved many hundreds of lives and almost certainly prevented a greater catastrophe during the final sad days of the RPA's siege of Kibeho. My message is simple and plain: that peacekeeping may be much harder than can ever be expected because it is a form of operational duty that invariably involves dangerous and unpredictable people. Accordingly, such operations may be very stressful and a number of supporting military, UN and non-government organisation personnel will require professional and timely stress debriefing as a result.

Without doubt, Rwanda was a crazy place of never-ending contrasts, and I am not surprised that some ADF members are having difficulties coming to grips with their experiences. I was privileged and indeed humbled by my experience of working with so many fine members from the three services, including their reserves. One of the successes of this conference will be measured by how well we care for our Rwandan veterans in future—and that success will be measured by what collaborative, proactive steps we take to prevent severe stress reactions and PTSD in future deployments.

What of the future? I would like to be able to identify those at risk of severe stress reactions and perhaps PTSD. I would like to see aggressive prevention programs and timely, active interventions instituted in the ADF. Equally important is our need for a comprehensive and effective method of national follow-up. Such a program would allow us to give the most appropriate, compassionate and efficient support to patients. It is encouraging that the ADF has been conducting a study designed to understand the effect of exposure to traumatic events on personnel deployed to Rwanda. It should, however, be noted that the study is being conducted as part of normal follow-up procedure that we have in place for ADF members after an overseas deployment.

This study has analysed the 311 ADF members deployed to Rwanda in the second contingent. It has examined the events they experienced and the level of distress that personnel underwent between 14 months and two years after their return to Australia. The study has also investigated the role of social support in the development of maladaptive traumatic stress reactions. Several demographic characteristics of the group were measured, including gender, marital status and type of service and rank.

The results of the Rwanda study indicate that all individuals that responded to the survey experienced at least one traumatic event: 79 per cent of respondents believed that they were in danger of being killed at some point during the deployment. At follow-up, some 20 per cent indicated that they were suffering distress, but this is not significantly different from the general Army sample. Other findings suggest strongly that the factors of social isolation and social support play an important role in the development of PTSD symptoms. Of particular interest is the fact that high scores on the loneliness scale are predictive of PTSD symptomatology at 12 months.

THE CASE OF THE UNITED NATIONS INTERIM FORCE IN LEBANON (UNIFIL)

The second case study I would like to refer to concerns UNIFIL in the Middle East. UNIFIL was established in southern Lebanon in March 1978 to monitor the withdrawal of Israeli forces, to help restore international peace and security, and to assist the Lebanese Government in restoring sovereignty over the area. However, UNIFIL's operations were hampered from the outset by a lack of consensus between Israel, the Palestine Liberation Organisation (PLO) and Syria.

The large Norwegian contribution to UNIFIL resulted in a comprehensive research project on acute stress reaction and PTSD being conducted by Professor Lars Weisaeth. In 1980, Professor Weisaeth served with UNIFIL as battalion medical officer to the Norwegian contingent. His study covered nearly 16 000 subjects of the first 26 Norwegian contingents to UNIFIL in the period between 1978 and 1991. The Weisaeth study uncovered that Norwegian military personnel had suffered considerable amounts of stress of specific types, but comparatively small amounts of the more classic combat trauma. The health of Norwegian personnel before, during and after UN service was generally good. However, 5 per cent of the sample had post-traumatic symptoms at follow-up on average at 6.6 years after the end of their service.

During 13 years only 530 soldiers, or 3 per cent of all Norwegian military personnel, were repatriated. These numbers were fairly evenly divided into three categories, covering illness and injury, disciplinary reasons, and social and family problems. In all, 96 per cent were satisfied with their experiences during their service; 90 per cent reported that their UNIFIL service was meaningful and that it had contributed to their positive personal development as well as conferring valuable military experience. However, those members that were repatriated short of a full tour of duty experienced more unfavourable outcomes. The premature termination of service was clearly accompanied by a feeling of failure and incompetence and by the need both to conceal and to rationalise the true reasons.

The Weisaeth study showed that early repatriation could be predicted by poorer quality of home life during adolescence, by higher prevalence of negative life events before entering service, and by a pronounced introverted personality. Of this subsample, 15 per cent had post-traumatic stress symptoms, a rate three times higher than other categories. In addition, an increase in alcohol consumption was very common for all groups, and a substantial number were unable to revert to a lower consumption level on return to Norway. While the total mortality during the observation period was lower than a similar cohort in the civil population, accident and suicide deaths were more frequent.

Of the study group, 60 per cent suffered violent deaths, including suicides, compared with 40 per cent of a matched group of soldiers that had not served in UNIFIL or on other peacekeeping missions. Weisaeth examined closely whether this was due to some selection factor, such as excitement seekers volunteering for overseas service, or whether it could be attributed to some factor during their peacekeeping tour, such as traumatic exposure. He concluded that traumatic events and secondary post-service adjustment factors mostly influenced higher mortality. The high violent death rate is a disturbing conclusion, which we in the ADF will need to follow up among our own veterans in future.

RECENT ADF COUNTER-STRESS INITIATIVES: THE COMMANDER'S GUIDE AND THE COLLABORATIVE COURSE ON STRESS SYNDROMES

Two recent and important ADF initiatives are the development of a commander's guide on operational stress management and the creation of a collaborative course on stress syndromes. The commander's guide was designed to demystify the term stress and to present information in a straightforward manner. The document covers the sources of stress while emphasising prevention and post-deployment management as well as training. The guide has been well received throughout the ADF and by other organisations, including state police services, fire fighters and state emergency services. In addition, our US, British and Canadian partners have warmly welcomed its contents, and the document has been adopted by the Department of Peacekeeping Operations in the United Nations.

The third ADF initiative is the development two years ago of a collaborative course on traumatic stress syndromes. This course has been developed jointly by the Defence Force, the Department of Veterans' Affairs (DVA) and the National Centre for War-related PTSD. Its aim is to educate participants about the nature of traumatic stress syndromes and to explain their relationship to trauma consequences and prevention strategies. The approach is multidisciplinary, involving health service officers, psychologists, social workers and chaplains. The course contains discussions on the nature of trauma, clinical presentation of acute stress disorder and PTSD, as well as considerations of prisoners of war (PWs), the experinces of torture survivors and assessment of hostage situations.

The key to the success of this course was that it comprised a practical phase in which the attendees were able to interview and interact with patients suffering from PTSD. Another key feature was a section on the commander's perspective. The latter included presentations from former commanders of peacekeeping forces and from Commander Australian Theatre (COMAST). A second course was held at the National Centre and appears to have been very successful. It is an excellent example of what can be achieved through close collaboration between the ADF and DVA.

POST-TRAUMATIC STRESS AND FUTURE THREATS TO THE ADF

Operational stress is unlikely to diminish in the future. The threats that will be posed to the ADF will probably be three-fold: operational, environmental, and occupational.

Operational threats are threats posed by warfare systems likely to be used by potential adversaries, such as small arms, mines and bombs. In future, this arsenal is likely to expand to include new and advanced armour-defeating weapons, incendiary devices and volumetric weapons. There may be, under certain circumstances, greater risks from chemical, biological and radiation weapons. Operational threats may also include non-lethal weapons. Such weapons are explicitly designed and employed to incapacitate personnel or materiel temporarily, while minimising fatalities and undesired damage to property and the environment. It is precisely these characteristics that make non-lethal weapons particularly attractive in peacekeeping operations. Non-lethal weapons include stun grenades, pyrotechnic flash devices, laser dazzle devices, tear gases, psychoactive compounds and foam barriers.

Environmental threats—that is, threats posed to ADF personnel by elements in the natural environment—include communicable diseases and toxic hazards. In the 21st century soldiers may be vulnerable to both new and old diseases, such as Ebola-type viruses and forms of re-emerging malaria. While communicable diseases pose the greatest environmental threat, ADF personnel are also likely to be exposed to a range of other environmental hazards including pollution, contaminated water, climatic extremes, ultra-violet radiation and toxic industrial chemicals.

Occupational threats—that is, man-made threats posed by our own warfare systems and equipment—include mechanical, electrical and thermal hazards, explosives, carcinogens, radiation and acoustic energy.

The most significant psychosocial threat is probably sensory overload resulting from the operation of complex communications systems and other information systems. We must ensure that the ability of personnel to assimilate knowledge is not overwhelmed by the magnitude of information to be processed.

CONCLUSION

This chapter has tried to demonstrate that peacekeeping may be more stressful than is commonly assumed by many members of civilian society. Examples from the Australian experience in Rwanda and the Norwegian experience in Lebanon provide solid evidence of the stress involved in deployment in peace operations. This is unlikely to change in the future—indeed, it may grow, and ADF personnel will continue to suffer acute stress reactions and PTSD as a result of both peacekeeping and warfighting operations.

While the ADF has a number of positive programs in place— the Rwanda study, the commander's guide and the collaborative stress syndromes scheme—there is still a need for more research and training programs to improve knowledge in the field of stress management. A key requirement for the future will be to standardise and integrate ADF psychology and human-science research functions to prevent, follow up and manage operational stress. Currently, the Australian Department of Defence does not have a centrally directed and well-coordinated human science research program. The field is fragmented between the Defence Science and Technology Organisation (DSTO), the Defence Food Science Centre (DFSC), the Defence Psychology Service (DPS) and various single-service agencies. The ADF requires a single human-science research agency along the lines of the British Defence Research Agency Centre for Human Sciences and the Canadian Defence and Civil Institute for Environmental Medicine. A centralised agency would be a key advance in helping the ADF learn from the experiences of peacekeeping operations in Somalia, Rwanda and East Timor.

6 | Balancing deep and close battle: will we still close with the enemy in the future?

John English

IT WOULD BE easy to conclude from a not too selective reading of the contemporary literature that modern technology and the growth of military professionalism have rendered conventional close-combat encounters redundant. The demands of forecasting the future face of battle have led many pundits to concentrate on what is now different in warfare, rather than what has remained the same. Of course, I would be among the first to agree with the practical statement of British tank theorist Major General J.F.C. Fuller that 'The object of all military training is to prepare the soldier for the next war, for it is the only possible war in which he can fight'. But I am also aware that many armies throughout history have failed to accurately determine the nature of the wars they embarked on, which Carl von Clausewitz considered the first and most comprehensive of all strategic questions. Obviously, to predict the future of warfare may be as difficult as picking the right investment in the stock market. In both cases we would probably be judged foolish if we did not pay some attention to past performance.

THE CONCEPT OF DEEP BATTLE

The concept of deep battle is not a new one, and the idea can be traced back to the 1920s and 30s. As early as 1926, Fuller had written that:

97

At present aviation can attack the enemy rear; tanks can break through the front and attack the rear; armoured cars can turn his flank and once again attack the rear . . . The attack of the rear at the present time is quite possible and in my opinion has become one of the most important tactical operations in war.[1]

The 1929 publication of Soviet General V.K. Triandfillovs, *The Nature of the Operations of Modern Armies*, also set forth a comprehensive theory for the conduct of successive deep operations. The object of these operations was to bring about the destruction of enemy forces throughout the depth of their deployments by means of breakthoughs and encirclements effected largely by long-range tank forces. By 1933 the Red Army had officially sanctioned as tenets of Soviet military art the concept of deep battle and—after the publication of *Field Regulations* (1936), prepared under Marshal M.N. Tukachevsky—the concept of deep operations.

During the Cold War, Marshal N.V. Ogarkov, Chief of the Soviet General Staff 1977–84, built on these concepts. In reviewing the growing size of armies since the 1500s, he came to the conclusion that increasing numerical strength had progressively expanded the spatial scope of military actions. The formidable accuracy and extended ranges of emergent conventional weaponry further convinced him that older forms of front (NATO army group) operations had to be replaced by theatre 'strategic operations'. By integrating long-range, terminally guided missile and air systems into automated reconnaissance–strike complexes (RUK in Russian), the whole territory of a country could be attacked. In Ogarkov's view, the enhanced mobility of armoured forces, attack helicopters and airmobile formations additionally augured operations being conducted over long distances at high speed by combined arms formations. Advanced guards, forward detachments and higher-level operational manoeuvre groups (OMG) resembling the mechanised and tank corps of the 1930s were thus expected to play an increasingly important role.

In 1984 Ogarkov went on to argue that nuclear stalemate had shifted the decisive role in war to conventional forces equipped with advanced technologies. In fact, as enemy field forces could be destroyed more efficiently through the employment of precision-guided and improved conventional munitions, there was no need to rely on nuclear systems at all. This development, which promised to enable higher commanders to intervene with air and missile strikes in enemy operational and strategic depth, amounted to nothing less than a military revolution in Ogarkov's estimation. As a defender armed with such military technology could conceivably wrest the initiative from an attacker by striking back quickly throughout the latter's depth, he also thought it likely to be the single most decisive factor in combat.

THE NATURE OF FUTURE BATTLE

What we must do, if we are to understand it, is to examine future battle through the prism of the past, for apart from training exercises and war games, which are only 25 per cent reliable, we really have little else other than historical research and experience to go on. Moreover, some of the biggest conundrums related to battle seem always to have hinged on the issue of continuity versus change. The great difference today is that technology appears to be driving change at breakneck speed, to the point where some argue that an entirely new military paradigm has appeared. We now talk of a transparent battlefield made possible by satellite surveillance, of doctrinal concepts such as network-centric warfare, and even of *creating* the military future in godlike fashion. Some also speak of improved communications heralding the return of the general on the white horse on the hill exercising effective command and control over battlespace. In this world of perfect situational awareness, it has further been suggested, the main role of conventional armies will be to support air power, to act primarily as 'beaters' flushing out enemy forces so that they can be destroyed by long-range precision fire. Yet, even in this futuristic scenario, one can discern a continuity reaching back to the Great War, where artillery once 'conquered' and infantry 'occupied', and to the Pentomic era in which the army role was to exploit nuclear strikes through armoured manoeuvre. That both of these concepts proved ephemeral should give us pause to think that perhaps future warfare may not actually unfold along the foregoing lines. We have to ask ourselves: what really has changed in war?

Lessons from the past

We know, of course, that we are not the first soldiers to wrestle with the problem of the nature of future warfare. We also know that war has assumed different forms in different ages, which would

seem to indicate that it is less a technological than a social phenomenon, perhaps still best described by Clausewitz, as:

> more than a true chameleon that slightly adapts its characteristics
> to a given case. As a total phenomenon its dominant tendencies
> always make a remarkable trinity—composed of primordial
> violence, hatred, and enmity, which are to be regarded as a blind
> natural force; of the play of chance and probability within which
> the creative spirit is free to roam; and of its element of
> subordination, as an instrument of policy, which makes it subject
> to reason alone.[2]

That war itself may wear the human face so easily discernible in
Clausewitz's matchless description could indeed explain why it has
periodically transmutated over the years. This, in turn, returns us
to the question of: where are we now? Has the age of the
mass-conscript conventional army truly passed once again and the
era of the long-service *armée de métier* arrived? Has Clausewitzian
inter-state war really been superseded by what Martin van Creveld
termed 'non-trinitarian' low-intensity conflict akin to that prevalent
in Europe before the Peace of Westphalia in 1648? Ironically, van
Creveld had no sooner made this argument in his book, *The
Transformation of War*, than the largest inter-state conflict since
Korea broke out in the Persian Gulf.[3]

In a review of van Creveld's book, Sir Michael Howard noted
that historians are as liable as anyone else to seize upon an ephemeral trend and project it into the future. Of course, van Creveld
could still be right, that future wars will be fought by guerillas and
terrorists rather than by armies, but neither he nor anyone else can
be certain. Surely it would be unwise to assume, as did the British
between 1919 and 1932, that there will be no 'great' (as opposed
to small) war for 10 years. Neither can one reasonably expect an
army trained chiefly for operations other than war to be as proficient in warfighting as one trained for war. This, too, was a lesson
learnt by the British Army, which up to the eve of World War II
was little more than a colonial gendarmerie charged with garrison-
ing and protecting the Empire. Though often engaged in internal
security operations and small wars against second-class enemies, it
remained better suited to imperial policing than to conducting
modern warfare on a continental scale.

As for van Creveld's assertion that guerilla warfare might well
be the wave of the future, the more one examines its modern
revolutionary variants the more one can see the critical significance

of the Maoist third conventional stage. The Civil War in China was won in 1949 by the huge conventional forces of the People's Liberation Army. Likewise, North Vietnamese regular soldiers and tanks proved decisive in the Vietnamese War. Perhaps the biggest mistake Western forces make is to get decisively engaged in such conflicts during the second stage, the guerilla stage proper, when more effective preventive action could be taken during the first stage, in the rough proportion of 75 per cent administration and 25 per cent fighting. In short, the best way to deal with potential guerilla wars in the future may be to head them off, through socioeconomic rather than mainly military measures, before they get seriously started. The second best may be to wait and destroy them conventionally in stage three.

In any case, it is perhaps premature to rule out the chance of inter-state conventional war occurring again in some unexpected circumstance or quarter. The Falklands and Gulf wars took the world by surprise, and there is no reason to believe that we will not be so taken again. It is also ironic that in the former case strategy turned once more on the marching pace of infantry, while in the latter the number of divisions still counted as a factor. The unique feature of the Gulf was that ground operations were contingent on, and clearly secondary to, the aerial campaign, which brings us back to our futuristic scenario and consideration of the close and deep battle. Here again, we should perhaps remind ourselves that the air offensives carried out against Germany and Japan during World War II also amounted to waging deep battle at the strategic level. The difference today is that theatre, army, corps and division deep battles have been more formally included in this mainly aerial calculus largely because of the increased strike range of available weapons. At corps and division levels, top artillery commanders are usually charged with coordinating the land–air battle in depth, up to the fire support coordination line (FSCL) in the case of division and beyond the FSCL in the case of corps using battlefield air interdiction (BAI), army helicopter aviation, and army tactical missile systems (ATACMS). One has to ask, however, whether this is conceptually any different from what corps and divisional artillery commanders did in conducting critically important counter-battery operations in World War II (and even in the Great War, where gas on a grand scale was used for this purpose).

The problem with using the Gulf War as a model is that it has fewer lessons to offer from the perspective of close battle. Attacking coalition forces enjoyed an unprecedented observational advantage wrought by an array of space satellites and advanced radar-surveillance systems. Unlike the Warsaw Pact, which possessed a satellite reconnaissance capability, the Iraqis remained strategically blind. The Soviet decision to deny them space intelligence, which might have been forthcoming during the Cold War, furthermore served to keep them in abject darkness. Had the Iraqis discovered the 200-mile westward flanking movement of Third US Army, they just might have been able to put up a better fight. As it was, coalition employment of high-technology means to destroy enemy air-defence systems, command and control facilities and artillery command posts rendered the Iraqis tactically and operationally, as well as strategically, blind. Without observation or intelligence, they were sitting ducks, unable to use their improved longer-range artillery. Indeed, it may be that no other field army in history ever fought at such a distinct disadvantage.

Neither was the Iraqi army, after eight years of fighting the 'longest conventional war of this century'[4], all that it was cracked up to be. The much-vaunted triangular tactical defence employed with great effect against Iranian hordes crumbled in the sands of Kuwait for lack of artillery and air support. Like the hapless Italian army of an earlier war in the Western Desert, the Iraqi army became a hostage to position in a treeless expanse of wasteland. Unable to negate enemy air action by hugging coalition ground forces or drawing them into close combat, as may have been disastrously attempted at Khafji, the totally exposed and ragged Iraqi defence could only receive blows, not deal them. In the end, the attacking divisions of VII Corps drove forward as an armoured phalanx, crushing all Iraqi forces before them. Having manoeuvred on a grand scale (which could not have been done in Europe), there was simply no need or even point to doing so at lower levels. Highly trained coalition tank crews also regularly scored first-round hits using thermal-imaging sights.

Had the Gulf War been fought in closer country with shorter sighting distances, coalition tanks would, of course, not generally have be able to stand off and safely score hits on enemy tanks from afar. In West Germany, where fully 55 per cent of sighting distances lay under 500 metres and only 10 per cent over 2000 metres, this could not possibly have happened, especially against a more evenly matched and better-motivated enemy possessing air and space-based intelligence and combat assets. One also has to remember that a major lesson of World War II was that experience gained in the Desert War was not necessarily applicable to Italy or north-western Europe.

Even though major coalition forces fought the Gulf War largely with equipment and warfighting concepts designed for NATO, it would be a mistake to assume that such a one-sided contest put them fully to the test.[5] The conclusion that the same outcome would have occurred on the Central Front had NATO and the Warsaw Pact clashed seems also to defy critical analysis. It was one thing to defeat a small nation like Iraq, another to fight a major power like the Soviet Union. Extrapolating lessons from the Gulf War within the context of the NATO–Warsaw Pact confrontation might therefore render them more tenable. In other words, to study the unfought Cold War between powers of the first rank may well be a better guide to understanding future warfare than examining the Gulf War in isolation. It may also be one of the few instances where preparing for the last war might be wiser than attempting to fathom the shape of some future imaginary conflict.[6]

The Central Front confrontation still has something to teach us about the close and deep battle relationship. In fact, NATO specifically introduced the latter to redress the problem of the former. Of the eight national corps that faced the Warsaw Pact, only six—three German, two US and one British—were ever considered first-rate. Even then, the bulk of their personnel were stationed more than 100 kilometres away from their general defence areas (GDPs). The most serious situation involved I Belgian Corps, which was responsible for defending the critical southern flank of Northern Army Group (NORTHAG) adjacent to Central Army Group's (CENTAG) III German Corps. Although 80 per cent of I Belgian Corps, including the majority of corps and logistics troops, occupied barracks in Germany, only one of its two-brigade divisions lay east of the Rhine. By most estimates, only two-thirds of a division equivalent would have been available to defend the I Belgian Corps sector seven days after any Warsaw Pact mobilisation.

Moreover, had the Soviets smashed into I Belgian Corps around 1982, they would have been opposed by no more than 1500 infantry, arguably still the only arm capable of holding ground. With but 750 foot soldiers in each division and 336 per brigade, any viable defence of the Belgian sector by day and night would simply have been impossible. To Belgian General Robert Close, the situation was 'almost grotesque', as such numbers, never more than 100 men per square kilometre, could not support anything more than a 'pearl necklace defence'.[7] In marked contrast to a Belgian division, just one of three motor-rifle regiments in a 13 000-man Soviet motor-rifle division *dismounted* 756 infantrymen. I might add that the West Germans also remained concerned that a *panzergrenadier* platoon dismounted but 18 infantrymen; a company, 70; and a battalion, 210—as opposed to a Soviet BMP regiment, in which a platoon dismounted 23; a company, 81; a battalion, 245; and a regiment, 756 infantrymen.[8]

Unfortunately, the tactical significance of such a human disparity often escaped the notice of observers intent on examining weapons systems balances rather than the actual nature of combined-arms operations. Plainly, the importance of infantry has not always been recognised, particularly in an age of machines, but one of the supreme ironies of history is that the lowly fighting man on foot has remained an absolutely indispensable factor in war. Using the soldier as the basic unit of mass and rough measure of an army's overall strength has thus continued to work reasonably well.[9] Within modern field forces, moreover, every additional tank and tank destroyer, if not necessarily headquarters filled to over-flowing with signallers and cipher clerks, really requires more rather than fewer infantry to provide intimate support and all-round protection in fair weather or foul.[10] The inescapable fact also remains that, as infantry incurs roughly 70 per cent of all battle casualties, *any* division that loses more than 50 per cent of its infantry component is, for all intents and purposes, in need of reconstitution. Headquarters and 'tail' services may still continue to function efficiently, but the divisional blade will have lost its cutting edge.[11]

Although fielding sufficient fighting formations remained the fundamental NATO problem on the Central Front, the course finally embarked on by the Alliance betrayed more of a techno-logical than operational orientation. In 1979 General Bernard W. Rogers, Supreme Allied Commander Europe (SACEUR), directed his headquarters to examine the use of proven and promising conventional weapons systems to trim Warsaw Pact attack elements down to sizes that could be handled by NATO forces deployed in GDP positions. This eventually resulted in the production of a document entitled 'Long Term Planning Guidelines for Follow-on Forces Attack' (FOFA), which on the formal approval of the Defence Planning Committee in November 1984 introduced a new

NATO theatre-level operational subconcept designed to raise the nuclear threshold.[12]

While Rogers remained personally confident that NATO corps could handle the Warsaw Pact first operational echelon, he feared that subsequent second-echelon forces, reserves, and operational manoeuvre groups (OMGs) supported by massive air strikes, would prove too much for them. NATO's inability to sustain its forces adequately with trained manpower, ammunition and war reserve materiel would, in turn, have left him with no option but to request nuclear release within a few days. On the other hand, the simultaneous disruption and destruction of Warsaw Pact rearward forces by conventional means up to 300 kilometres in their depth offered hope of shoring up forward defence while reducing dependence on nuclear interdiction. This was to be accomplished by FOFA, coordinated at the AFCENT–NORTHAG–CENTAG joint operational level,[13] initially through the application of proven technologies such as manned aircraft, missiles with conventional warheads, cratering munitions and electronic warfare. Unlike Soviet deep-battle concepts, FOFA did not involve airmobile or other ground forces in cross-border operations. Neither did it entirely replace NATO's dated option of employing nuclear weapons for deep strikes on the battlefield.[14]

The object of FOFA was to delay, disrupt and destroy follow-on forces from the front line to as far into the enemy rear as NATO target acquisition systems and conventional weapons reached. As the Warsaw Pact possessed formidable air defence assets, covering even the movement of OMGs in the first operational echelon, the subconcept also called for the suppression of enemy air defences (SEAD), attacking enemy airfields, and gaining air supremacy. The development of unmanned and stand-off delivery means also promised to reduce high loss rates expected from the deep penetration of Warsaw Pact airspace by all too few NATO-manned aircraft. By using a variety of new and old aerial means to attack enemy follow-on forces at times and places where they were most vulnerable, NATO hoped to retard the forward movement of Warsaw Pact reinforcements and thereby reduce the overwhelming momentum of the offensive. Given the finite number of attack corridors available to Warsaw Pact armoured forces, NATO planners envisaged acquiring lucrative targets at critical choke points such as railheads, airfields and river-crossing sites, where troop concentrations would most easily have been detected. As FOFA did not involve the use of nuclear weaponry, of course, formal release procedures were not required to initiate such strikes.[15]

Whether or not FOFA would have worked remains a matter of conjecture. To some it was a clever way of achieving depth that was politically difficult to obtain within West Germany, but it was hardly true depth. In its essential features the subconcept amounted to little more than aerial attrition in the enemy rear, as compared to Soviet ground thrusts with OMGs. It carried the complexity of the battlefield to new heights. The numerous functions comprising FOFA included surveillance to obtain information about the location and activities of potential targets; processing such data into usable information and transmitting it to appropriate commanders; deciding which targets to attack and relaying all necessary information to the attackers; conducting a more detailed reconnaissance to verify and track selected targets; delivering a weapon onto the target in accordance with the latest information; and assessing the results of individual strikes. To target *mobile* Warsaw Pact formations additionally required the deployment of more advanced high-technology systems, such as the US Army–USAF Joint Surveillance and Target Attack Radar System (JSTARS) used with such success in the Gulf.[16]

Critics of FOFA contended that steep investment costs associated with the development of high-technology systems risked diverting scarce resources from NATO's efforts to improve standard conventional defences.[17] Just as nuclear deterrence tended to focus military attention on the grand strategic plane to the neglect of operational and tactical dimensions, too much preoccupation with fighting a deep aerial battle further risked compromising the still vital close-contact struggle against enemy first-echelon forces (which, in fact, called for striking shallow rather than deep). The requirement to verify exact locations of tracked moving ground targets exercised a further drag on the efficacy of aerial engagements. In addition, the Soviets possessed both surveillance and anti-satellite (ASAT) systems as well as long-range surface-to-air missiles (SAMs) capable of shooting down airborne warning and control system (AWACS). JSTARS, which comprised aircraft carrying ground-imaging radar and extensive processing equipment plus a communications network and ground stations, had also to range well back from the forward line of troops (FLOT) for security.[18] As the Germans deemed the first-echelon threat to be the most dangerous and strikes 150 kilometres beyond FLOT not cost-effective, they opted to make the close-battle and air superiority missions their top two priorities.[19]

CONCLUSION

Obviously, at the operational and tactical levels there is a direct connection between the close and deep battle, but the purpose of the last must surely be to facilitate the first. As we learnt at the strategic level, the Douhetian battle-in-depth—whether effected by heavy bombers or missiles delivered from the air, as in the case of the Gulf—has not shown that it can by itself break an enemy's will to fight. To expect operational and tactical aerial attack in depth to exert war-winning leverage on its own is therefore equally questionable. Besides, as Soviet practice and theory have indicated, the battle-in-depth can also be waged by ground forces that must necessarily be developed from the close battle. Indeed, in the final analysis we are really talking about one battle.

The unique circumstances of any given situation will also continue to dictate how battle unfolds. Under conditions of nuclear threat or precision aerial attack, troop safety may well lie in closing with the enemy. This is, after all, why German troops in World War II often crossed to the 'friendly' side of Allied bomb lines and why Soviet infantry at Stalingrad and the PLA in Korea resorted to 'hugging' tactics. To a lesser extent, this was also the rationale behind OMG operations. Whatever the case, the close battle in urban and forest fighting is likely to be as chaotic and casualty-intensive as it ever was—if only because 'noise' has now been added to the friction and fog of war. Moreover, because combat at this level has always been characterised as a series of local actions, small groups can probably be expected to play the role they always have. They may even become more important as port seizure or denial assumes increasing significance.

The importance of ground, of course, remains inextricably linked with *observation*, which is still the key to success in war. There is a lot of truth in the old saw that only an invisible weapon can do its job properly. No conventional army can expect to survive if it can be observed with impunity by an enemy possessing a god's-eye view. In such circumstances, spy satellites and other intrusive surveillance systems must either be destroyed or blinded through passive measures that, in the future, might include improved battlefield fog or anti-thermal smoke. Failing this, soldiers will have to seek out the protective cover of urban terrain, develop practical 'stealth' means themselves, or rely on their wits to deceive. In truth, however, their task would be made much less daunting by the mere existence of a situation of air parity.

On that note we come back full circle to the not-to-be-discounted scenario of a war between roughly equal opponents, which would bring back all the traditional challenges of, *inter alia*, how to deal with modern defensive systems and why force ratios still remain important.

7 | Fighting spirit: leadership and morale on the 'empty battlefield' of the future

Paddy Griffith

THIS CHAPTER AIMS to consider the chaos and stress of a future battlefield, and how they may appear to a member of the Australian infantry who finds himself (or perhaps even herself?) exposed to them. Having identified a few of the salient psychological problems, I will suggest some ways in which 'fighting spirit, leadership and morale' might best be developed. As a preliminary, however, a warning should be issued that history is a much more certain science than fortune-telling, and I would acknowledge some unfortunate brushes with the latter activity.[1]

EXTRAPOLATING FROM THE PAST

If we do have to look into the future, we should at least approach it by extrapolating existing trends forward from the past, rather than imagining that some completely revolutionary 'brave new world' will suddenly break completely free from everything that we already know. The British and Commonwealth tacticians fell into this trap in 1941, when they imagined that the war in North Africa would be an entirely novel and 'pure' experience, with no terrain, no civilians, no infantry or artillery, but merely the tanks cruising around for fleet engagements like battleships at sea. In this illusion they were strongly encouraged by the characteristically misleading prewar writings of J.F.C. Fuller, and it would take some 18 months of defeat and humiliation for them to see the error of their ways. Then, with the arrival of Montgomery in the autumn of 1942, they finally reverted to a system of combat that was essentially unchanged from World War I. It was not at all futuristic, and was even boringly unimaginative; but it did at least produce victories, whereas visionary theorising had failed very dismally.[2]

What, then, can we expect the battlefield to be like when our future Australian infantryperson steps out onto it? In the first place, we may suggest that it is unlikely to be particularly chaotic, or difficult, or dangerous at all, at least nine times out of 10. Given the balance of power that is likely to obtain in the foreseeable future, we may surely assume that, if Australia deploys any military force, it will be a relatively small but highly professional contingent, and it will confront an enemy that it will be able to handle with relative ease. The task may be essentially a peacekeeping or policing one, in which serious combat is unlikely to occur; or perhaps it will be a 'military promenade', like the Gulf War of 1991, in which technological superiority will weigh so devastatingly in favour of the 'First World' side that the fighting will be gratifyingly one-sided. Probably the Australian contingent will be able to draw support from a much wider international force, as has so often been the case in the past; but even if it is acting alone, we may assume that it will bite off only what it can chew.

If she fails to exercise such caution, Australia may indeed find herself involved in a national emergency, where a hasty mass mobilisation will be necessary, and where undertrained 'civilians in uniform' may have to be thrown into the fighting unprepared— as they were in the Owen Stanleys in 1942. However, that will make a very different scenario from the type that we are discussing here, and will require some very different mental and theoretical preparations from the ones that will be described.

If Australian regular troops are deployed properly, it is likely that they will enjoy a superiority over the enemy in training, physical fitness, weapons, mobility, communications, battlefield surveillance, logistics, and most other branches one might care to name apart, perhaps, from numerical strength. They will expect to have the upper hand in most circumstances—and let us hope that they do so in all circumstances. They will therefore face little stress, chaos or real danger. War, however, is a notoriously chancy business, and so they must always be prepared for that one time in 10 when they will *not* find themselves enjoying such an easy ride. There is always a possibility that chaos, stress and high levels of danger may unexpectedly rush back to centre stage. This may occur if something goes badly wrong with the deployment, or its underlying political basis is misjudged, or if communications crash, or if some error of leadership allows a hopelessly small force to be engaged by a very big one. The Battle of Long Tan in 1966 is one obvious Australian example of this. However, at a more strategic level, it may not be irrelevant to note that in April 1982 it was through a deliberate act of British government policy that just two companies of Royal Marines were left stranded on the Falkland Islands to face an invasion by the entire Argentinian Army, Navy and Air Force.[3]

Let us suppose, then, that through some blunder—or merely through the workings of chance and the unexpected—our Australian infantryperson does in fact suffer the misfortune of getting into serious combat. What will it be like? The most appropriate adjective will surely be 'bewildering', as no one person will be able to see more than a small part of the battle, and will probably not see more than two or three of their own side at any one time, let alone the enemy—and let alone knowing the intentions that lie behind what either of the two sides are trying to achieve.

If combat were easy for eyewitnesses to understand, then it would be equally easy to describe to others: but we know from the experience of many generations that this is not the case. At the battle of Bautzen in 1813 the novelist Stendhal, who happened to be present as a member of Napoleon's army, reported: 'I have seen everything that anyone can possibly see of a battle—that is, nothing'. Shots will fly around; loud noises and blast waves will shock the senses; men may fall and bleed and scream beside you; but you can never really expect to understand very much of what is going on, even if you have been given clear orders and there is relatively open visibility (neither of which conditions is likely to apply). Probably the orders will appear irrelevant and the lines of sight will be short. It is remarkable how often battles are fought in built-up areas or in jungles, forests or 'bocage' (hedgerow) country—and in extremely adverse climatic conditions. Arguably the pleasant summer of 1944 in the shady green meadows of Normandy represented the best sort of weather that any soldier could possibly wish for—but we cannot say that the campaign was reported in those terms. Instead, survivors remember the violent thunderstorms, the putrescent stink of dead cows, the buzzing of flies, and the lethal silence from which a hidden sniper might claim a victim

at any moment. In the front line the troops were not free to luxuri-
ate on the greensward, but had to cower in foxholes dug out of
the gritty soil. When their view of the battle was not restricted
to the earth and tree roots six inches in front of their faces, it
normally consisted of only a small corner of an empty field, as
seen from ground level.

THE 'EMPTY BATTLEFIELD'

This bewildering lack of information and visibility is often
described by the expression the 'empty battlefield', although that
is a rather ambiguous phrase. It can cover at least three different
factors—observation, dispersion and the range of weapons—and
therefore perhaps deserves a little clarification here.

First, it has been used to refer to the increasing use of cover,
camouflage and concealment during the 19th century, when soldiers abandoned their bright red coats in favour of khaki, and
started to hide in folds of the terrain or in trenches, instead of
standing up proudly in full view of the enemy. This trend had not
been completed even in World War I, as although the troops were
protected by trenches, the trenches themselves were usually fully
visible to the enemy. In World War II this was less often true,
and a much greater proportion of an army's positions were truly
hidden—including belts of mines, which depended for much of
their effect upon their concealment. Since 1945, however, great
strides have been made in devising new technologies to detect the
enemy in all weathers, or at night, through such signatures as his
smell, his electromagnetic emissions, his noise, the metallic density
of his vehicles and, above all, his infrared profile. Such futuristic
studies as the Australian Wundurra project now even allow the
infanteer to 'see round corners' in house-to-house combat.[4] Many
attempts have admittedly been made to produce effective methods
of cloaking or suppressing telltale signatures, and some remarkable
successes have already been achieved even with primitive technology. Nevertheless, this surely still remains an area in which
investment in higher technology lags woefully far behind the
potentially massive benefits that a genuine breakthrough might
bring. One remains unconvinced that attempts to update the
techniques of tactical camouflage have been particularly effective,
apart from a few specialist cases such as aircraft stealth technology
against radar, or plastic mines against metal detectors. Hence

I believe that we now have a good chance of seeing what there is on the battlefield, at least if we have leisure to make a careful analysis and follow the full 'military appreciation process'. To that extent the battlefield of the future will prove to be considerably less 'empty' than that of the past.

It is often claimed that this new visibility of the battlefield will inevitably mean that operations will be extended through the night, all night and every night, with a correspondingly dramatic increase in fatigue and stress. Physiologically it is indeed possible to keep up such 'continuous operations', after a fashion, for up to three or even five days without a break. However, it is not to be recommended, as performance significantly diminishes during the five days, and by the sixth day it starts to collapse altogether. Nor have we seen such operations attempted in practice, even by armies possessing modern night-vision equipment, for periods longer than that. Of course, throughout history it has always been possible to conduct operations at night—and today they are doubtless easier than ever—but that is not at all the same thing as continuing all day and all night over a long period of time. It has to be done in shifts, as it always has been done; or if you find you enjoy an advantage over the enemy at night, then it pays you to cash in on it—but to sleep soundly during the days in between.

Second, the 'empty battlefield' may refer not to difficulties of visibility but to the wide dispersion of troops and weapons in both breadth and depth. This began when the age-old habit of fighting in closely packed masses began to give way to looser formations of skirmishers. It may be traced at least as far back as the hedge fighters of the English Civil War in the 1640s—but perhaps it became most noticeable around 1914, when even the closest-packed battalion line was supposed to be only one rank deep, with an interval of several feet between each man. During the Great War the trend towards dispersion accelerated rapidly, until by 1917 the basic unit of tactics had become a small but heavily armed squad of skirmishers acting almost independently of higher control, and almost completely lacking in drill formations.[5] In World War II we even saw the concept of the 'one-man front' in jungle terrain such as the Kokoda Trail, although in the end it was found to be psychologically necessary to have at least two men in close touch with each other, so no-one would feel completely alone. In his reports on the fighting on some of the Pacific atolls, for example, S.L.A. Marshall suggested that the two-man crews of machine guns remained motivated for much longer than the riflemen who were dispersed as individuals, who were much more likely to lose sight and sound of their comrades.[6] In future each man will doubtless carry his own personal radio to link him to his squad commander, so to that extent he will once again be expected to operate out of sight, and 'act alone' as the basic unit of combat; but in human terms he will certainly still need at least one 'buddy' in close physical proximity. The two-, three- or four-man buddy group, and not the one-man bionic warrior, must therefore be considered the irreducible unit of combat, beyond which the battlefield may be emptied no further. In practice, the exact size of each team will doubtless normally depend on the numbers of crew required to man each vehicle.

In terms of a battalion's military effectiveness, we can suggest that the modern platoon is potentially little less powerful than a company was in Vietnam, and perhaps than a whole battalion was in 1945. It will certainly hold far more radios, just as each individual soldier's assault rifle may today be considered equivalent to the squad light machine guns of earlier times. A troop of tanks can probably achieve more 'first-round kills' on enemy tanks than an entire armoured division could manage in 1945, and at longer range. The continuing tendency will therefore be for a platoon to do a company's job, and a squad to do a platoon's, and hence for each junior commander to operate more independently and physically further away from his higher HQ. He will also find himself entrusted with a correspondingly heavier load of responsibility, and wider powers to make significant decisions. We are not far away from the mythical situation in which 'a corporal can fire a cruise missile'.

Against this we must remember that improvements in communications have historically always helped to strengthen centralised control and inhibit initiatives at lower levels, but the converse is also true. Improved signals enable wider dispersion and therefore necessarily increase the need for many types of independent decision-making. For every higher commander who tries to micro-manage his subordinates, there will surely also be a group of subordinates who selectively switch off their rear-link radios and thereby impose their own type of devolution on the situation. Then again, the multiplication of incoming message traffic can quickly lead to 'the fatal disease of information overload'[7] for a higher commander, unless he is willing to delegate responsibilities back to his subordinates. Hence decentralisation, or what has variously been called '*Auftragstactik*', 'mission orders' or 'network-centred warfare', will surely increasingly impose itself. We should beware of over-conceptualising this, or treating it as mainly a problem in the 'theology' of doctrinal development; but it remains true that it is growing, will continue to grow, and within reasonable limits should not be resisted.

Third, the idea of the 'empty battlefield' may be used in connection with neither the degree of camouflage nor the distances between men, but with the range at which fire is delivered. Wellington's infantry would wait until they could see the whites of their opponents' eyes—perhaps at only 30 yards—but from 1870 onwards it became normal to train infantry to open fire at ranges of up to a mile. In the battle of Isandhlwana in 1879 one essential fallacy in this concept was cruelly exposed when the British fired off all their ammunition at long range, leaving them defenceless when the surviving Zulus came to close quarters. However, in most other wars a much more important fallacy was frequently revealed, as it was found to be rare for the enemy to be visible at ranges of anything like a mile. In the two world wars a range of 300 yards was considered to be unusually long for rifle fire, and often it was much less. The design of assault rifles since 1945 has certainly veered away from seeking accuracy at long range, despite a long rearguard action fought by purists in the British Army. In Vietnam the average engagement range was nearer 10 metres than 20, which meant that the M16 made a logical successor to the Owen gun and the Mk V Lee Enfield jungle carbine, combining the most useful qualities of both.

Of course, throughout the 20th century many of the auxiliary weapons attached to the infantry have habitually operated at much longer ranges. The tripod-mounted machine gun, for example, could fire on fixed lines as far out as 4700 metres. Similar ranges applied to mortars, trench cannons and fire from supporting tanks or APCs; while in recent times not only have helicopters been able to deliver aimed fire within that envelope, but artillery has at last become sufficiently responsive to be called down by almost every individual soldier, and almost in single shots with a good chance of a one-shot hit. In theory all this means that the real range of infantry fighting has now definitively extended even beyond the 'one mile' claimed prematurely in the Zulu War; but in practice we can suggest that it has still not solved the problem of close country such as jungle or built-up areas. Especially in an unexpected engagement, where there has been little time for sensors to reveal the full secrets of the enemy's dispositions, we can expect most of the really important exchanges of fire to take place within the line of sight of front-line troops.

The enemy may still often present himself at very close range, even though the battlefield may otherwise remain 'empty'. For present purposes, therefore, let us use the phrase the 'empty battlefield' to suggest that the individual soldier can generally see little of what is going on, and knows little about what is in the minds of the two opposing commanders. He has to suffer all the horrific consequences while remaining ignorant of their inner meaning, and it is perhaps this ignorance, above all other factors, that can make combat feel so chaotic and stressful.

ESCALATING STRESS IN FUTURE WARS

It has often been suggested that the chaos and stress of future battles will inexorably rise to even higher levels than those of earlier wars, and that the future will be more horrible than the past. For example, when John Keegan wrote *The Face of Battle* in 1976, he seemed to believe that modern battle had become *so* stressful and *so* chaotic that it would be impossible to bear, and therefore it would 'abolish itself' and become extinct.[8] We may reply that it is surely humanly impossible to imagine anything worse than some of the fighting that occurred in the two world wars; nevertheless, we must admit that there are a number of factors that support Keegan's thesis.

Weaponry

The first factor promising to make battle more difficult is greatly improved weaponry. We have already mentioned some of the ways in which this may affect visibility, dispersion and range; however, we must also remember that many new weapons promise a dramatic increase in casualties among armies that are caught unprepared or simply behind the times. The 'one-shot kill' capability of many types of 'smart bombs', whether delivered from the air or the ground, has been widely discussed in the press; but it is worth remembering that their lethality derives not only from improved accuracy. Their heightened explosive, penetrative or fragmentation power, once they reach their target, is also an important part of their effectiveness. Air-to-ground fire is particularly to be feared, and not even the dreaded Stuka dive-bombers of

1940–43, or the Typhoon ground-attack aircraft of 1944–45, were anything like as dangerous to ground troops as a modern attack helicopter or A–10 anti-tank aircraft. Equally, mines have improved beyond all recognition since 1945, and have been joined by a large bestiary of other horrors, ranging from chemical and biological agents to lasers designed to strip the retina off anyone who so much as looks at the enemy. The unfortunate infantryman who finds himself facing a truly modern army may be fully justified in believing not only that he has no hope of survival but, still more stressfully, that he and his comrades will have no means of fighting back.

Demographic trends

Not only have modern weapons become distinctly nastier, but the people who will have to face them have perhaps become softer—effete 'couch potatoes' addicted to junk food, junk television, obsessive computer games and omnipresent airconditioning. They are unused to the hard physical conditions that our grandparents took for granted. In the particular case of Australia, moreover, recent demographic trends have led to a certain dilution of the supposedly 'martial races' that provided the original Anzacs. Many of the ethnic groups in modern Australia that are not of 'English-speaking background' (ESB) have been reluctant to volunteer for the defence forces, thereby leaving the ESBs as a sort of exclusive in-group. This implies that the pool from which recruits are drawn is considerably narrower than it might otherwise be—except, perhaps, in the case of major wars of national survival, in which mass conscription would be unavoidable.

Beyond the questionable fitness and willingness of the modern population to enlist, there is the still more awkward fact that most inhabitants of the developed world have been brought up to believe that they have an absolute right to live out a full term of at least 80 years, if not 100. People who fail to survive their 50s are today considered to be 'dying young', whereas a thousand years ago a Viking warrior would have had to die before puberty in order to earn such a description. Even in the Great War of 1914–18, the chance of death or serious injury of about one in five for those joining the forces was not seen as excessive for many classes of society, when set beside the chances expected in civilian life. The average life expectancy of a coal miner was no more than 45 years in 1914—and women in childbirth would suffer a similar hazard until at least the 1930s.[9] As a matter of fact, the chances of death in the Great War were not significantly higher per enlistment than they had been in any major war of earlier times, but what was new was that a much higher proportion of the population was enlisted (whether as volunteers or conscripts). The army took in large sections of society that would never previously have contemplated entering it, so naturally they were shocked and horrified by what they found. This effect has continued and multiplied ever since, as increases in material wealth have led to a huge expansion of the comfortable middle class, and a correspondingly dramatic reduction in the truly brutalised poor. The result is that we no longer regard death in the army as normal, or even as 'acceptable in special circumstances'. The general perspective of our liberal democracies, and of the mass media that speak for them, is today the same as that of the protesting war poets who found it so difficult to make themselves heard in 1917.

FACTORS COUNTERING GREATER STRESS IN FUTURE WARS

All is not lost for the possibility of future warfare, however, as there are still a number of factors that may act in the opposite direction, making future battle easier to bear. In the first place, the speed and efficiency of casualty evacuation has increased dramatically during the 20th century, especially with the introduction of the helicopter and the 'search and rescue culture' that seems to have grown up alongside it. A wounded soldier may now expect to survive, which was not at all the case in the Boer War or earlier, and only partially true during the two world wars.

As for the new technology of weaponry, it may paradoxically become less fearsome in direct proportion as it becomes more accurate and discriminating. Thus, the civilians living next to a vital bridge (e.g. in central Baghdad, 1991) may have less fear that they will be bombed flat—as so many Italian and French families were in 1944. The infantry platoon stationed next to one of Saddam Hussein's SAM batteries in the Iraqi no fly zone need not worry about the radiation-seeking missiles directed against their immediate neighbours. Of course, this is no consolation to the radar operators in the crosshairs of the incoming smart bombs, but at least the battlefield as a whole has lost some of its randomness. If it has become slightly more predictable, it is also likely to have become slightly less stressful. It is most predictable and least stressful when it is 'our' side that possesses all the best and most terrifying hardware (as has been true for the First World soldiers in most recent conflicts, with the new weaponry a help rather than a hindrance to 'our' side). So far had this process developed by 1991 that there were surely few real terrors for Western soldiers even in Saddam Hussein's gigantic tank army, and in the event it was dismantled with almost effortless disdain. It was doubtless a different story with his chemical and bacteriological weapons—but these should perhaps be considered at the operational level, rather than at the tactical level with which we are here concerned.[10]

As it happens, there was in fact one type of conventional tactic that had the potential to deter and even 'frighten' the Western planners in the Gulf War, although it was by no means new or technologically based. That was street fighting, or FIBUA, which might degenerate into a form of protracted guerrilla warfare. Quite apart from the diplomatic difficulties, the prospect of a bitter infantry struggle through the streets of Baghdad was politically unacceptable at the time, both because of the allied casualties that would have been incurred and for the devastation to innocent Iraqi society that would have been unavoidable. In that sort of combat the Iraqis would have been fighting much more on equal terms with the Western troops, just as the jungle had allowed the North Vietnamese to fight on equal terms with the Americans and their allies in the 1960s. The balance of human factors—the 'fighting spirit, leadership and morale' of the two sides—would have tended to become more important than small differences in weaponry.

When we turn to those human factors, however, we find that many of the pressures that tend to make the modern population 'soft' are in fact reversible, provided the process of reversal is started at a sufficiently early age. The potential soldiers must first be selected with care and only the 'right stuff' accepted. That is by no means a matter of selecting the toughest, strongest and bravest—nor the best sportsmen—but rather the most psychologically secure 'team players', who can be relied on to display genuine 'mateship' under extreme conditions. After that, the 'couch potato' may be slimmed down in the gym, and acclimatised to tough conditions by training exercises in the wild. Group cohesion can be built up wonderfully fast by a few months together in the jungle or the desert, while it is even possible to some extent to inoculate against squeamishness. Visiting hospitals, looking at photographs of destroyed bodies, and conducting surgical simulations with animal offal, can all prepare people to do their job without fainting at the sight of blood. A deep knowledge of medicine can be a great help, not only in practical first-aid for the wounded but in the psychological background preparation of the healthy.

FIGHTING SPIRIT, LEADERSHIP AND MORALE

One's soldiers have to be brutalised and tribalised, to some extent, if they are to operate effectively in the brutal, tribal world of combat. They will necessarily find themselves far removed from the comfortable niceties of middle-class society, and perhaps far more so mentally than physically.[11] Given time, the right sort of volunteers, and the right sort of political permission to set up a 'tribe' of this type, which does pose certain dangers to civilian society, the necessary transformation can be achieved. It should be noted, however, that we are here still talking about a small professional army, rather than a mass citizen army such as might have to be called up in an emergency.

Nor are we talking about brutal discipline within the tribe. On the contrary, it is logical that if you want your troops to be self-reliant, full of initiative and capable of operating complex technology in the interests of sophisticated tactics, you must trust them and win them over by reason, persuasion and leadership by example, rather than by harshness. This is precisely the logic that dominated the tactical debate of the later 19th century, when skirmishing was starting to take over from tactics in close order. Perhaps the best summary can be found in Ardant du Picq's *Battle Studies*, which were written in the 1860s (although they took time to become well known, as the author was killed in the Franco-Prussian War of 1870 before they had been published).[12] What Ardant said, in essence, was that new weapons were making combat ever more dangerous and stressful, and at ever-increasing ranges. Men felt more naked and vulnerable, and therefore more ready to abandon their training and tactics, to put a quick end to the danger by running either forwards or backwards—but in both cases running out of control by their officers. The solution that he recommended was to reduce this loss of control by fitting the tasks the soldiers had to perform more exactly to what their instincts told them to do. In this he indulged in some racial stereotyping to show that French soldiers under stress were naturally liable to disperse, whereas

Russians were likely to huddle together in a mass. He wanted to harness French excitability by abandoning harsh close-order drill and 'bull' and by allowing increased dispersion, but also providing drills and tactical insights to master and discipline the potential for confusion. He wanted to link this to a 'sociable moral coercion', by which all members of the group supervised and supported all the other members. The men were to be trusted more and bossed around less, because they would be individuals who were known to each other, and not just automata to be beaten into their places. He said: 'The remedy lies in an organisation which will establish cohesion by the mutual acquaintanceship of all . . .'.[13] In Prussia a few years later a dispersed skirmish line was to be controlled by the application of a very similar concept, called *Innerführung*, or the strengthening of personal moral qualities within each individual. This would replace the old draconian brand of 'external', close-order discipline that had been favoured by martinets from the school of Frederick the Great.

These principles were given practical expression during the Great War that was fought after, and to some extent as a result of, the debates about combat psychology that followed the life and work of Ardant du Picq. On the Western Front of 1917–18 a resourceful and professional force of skirmishers—or 'storm troops' who were self-motivating and capable of taking ground from the enemy—was developed in every army, and not least in the Australian Corps.[14] At their peak, this infantry became every bit as good as good infantry can possibly be, in any era of history—and its 'fighting spirit, leadership and morale' allowed it to weather the very worst storms of steel that the scientists could contrive, as well as the meteorological storms of wind, rain and sleet that made life in the trenches so notoriously uncomfortable. The Anzac storm troops of the Great War therefore offer us an excellent example of what it takes to create infantry that will stand up to extreme conditions of combat. It may be worthwhile suggesting a brief list of the significant factors as they appear to me, although doubtless Australian experts will wish to modify this list in the light of local perceptions:

- 'mateship' (not only sharing everything with your mates, but being desperate not to let them down);
- keeping together each 'tribe'—squad, platoon, company, battalion and so on—for a long time; preferably (as in 1916–18) all the way up to army corps level;

- a relatively informal hierarchy between officers and men (no 'bull'), and promotion from the ranks, purely on merit in the specific job of infanteering;
- physical resilience and toughness (including selection of recruits with a certain wild 'frontier' attitude), as well as plenty of physical training;
- plentiful training, battle-drill, practice, rehearsal, so not even the most unexpected things are truly unexpected; and
- complete disdain for the danger—and for the longevity—of the enemy; a culture of initiative and derring-do (even recklessness).

CONCLUSION

The aim is to put in place as much of a guarantee as is humanly possible that the organisation will continue to hang together, no matter what happens, and each individual within it will not only know what to do but will continue doing it as well as possible, regardless of mud, blood and bullets. If one wants to sum it all up in a phrase, 'unit cohesion' may be a polite way of putting it, while 'tribalism' is probably a little more accurate. Even if one does not favour that particular phrase, it should at least be stressed that the really important things have to do with psychology and social intangibles rather than hardware or technology. This historian's lesson for the future is therefore to continue to extrapolate from the past and quite simply to 'keep on doing what the Anzacs always have done'.

8 | The ghost of Jomini: the effects of digitisation on commanders and the workings of headquarters

Jim Wallace

ONE OF THE main motivations for digitisation in advanced Western military establishments is the belief that the process will impose a new order on the inherent chaos of the battlefield. While this is a doubtful proposition, it remains a powerful idea among many Western soldiers. It might be argued that there is an almost blind acceptance in some elements of the West's officer corps that the more digitisation there is, the better it will be for the profession of arms.

Yet an honest analysis of Western attempts at digitisation—from the US Army's Force XXI scheme, through the US Marine Corps' (USMC) Hunter Warrior program to the Australian Army's Restructuring the Army (RTA) trials—reveals a quite different picture. Collectively, these programs and trials demonstrate that digitisation has to be carefully applied to activities that may enhance the function of battle command. These activities include both the roles of commanders as individuals and their corporate embodiment in a headquarters.

Understanding the implications of digitisation requires a basic analysis of what is expected of Western commanders in the future as well as an assessment of the various environmental factors that can affect the success of command in war. In addition, it is important to grasp that digitisation is not a process that can be considered in isolation: its context must be clearly understood.

Digitisation is part of an increasingly complex communications and information revolution; it creates an environment that paradoxically can both harm and help a military commander.

DIGITISATION AND THE ROLE OF COMMANDERS

Leadership is the principal function of the military commander on the battlefield. It is true that the character of leadership may have changed to encompass both 'heroic' and 'directive' styles; but, like war itself, the nature of leadership remains unchanged. A form of heroic leadership remains necessary for the exercise of effective command at the platoon and company level and, perhaps occasionally, even at the battalion level. Today, however, most battalion and formation commanders tend to lead more by directive control through coordination of their subordinates' actions, while retaining their instinct for battlefield conditions. Senior commanders also retain the right to inject their knowledge and experience at critical points in the unfolding of military action. In reality, a form of directive control has been dominant in the history of war in the 20th century for most military clashes—except perhaps the very large-scale battle.

It is vital to understand that the unique challenge of battlefield leadership—whether manifested in a heroic or directive style—has not changed. The purpose of battlefield command remains the same as it was in the days of Alexander and Caesar: it is to motivate men and women to fulfil a hazardous mission in the face of personal costs that may defy reason and logic. Because the battlefield is untidy, unpredictable and chaotic, military leadership remains essentially a spiritual activity—a matter of both heart and soul. To lead successfully in battle requires the imposition of will and the transmission of firm purpose in a very personal manner. Battle leadership does not thrive under impersonal conditions for the simple reason that it is the epitome of personal communication.

The reality of both heroic leadership and directive control, the unpredictability of events, the reality of human weakness and personality, the issue of chaos and the world of the spirit are all at play in battlefield leadership. These multiple factors provide an essential background to understanding what digitisation may, or may not, achieve on the battlespace of the future.

THE LEADERSHIP ENVIRONMENT

It is important to determine the key aspects of the environment in which battle command is exercised. By doing this, it is possible to identify how the battlefield's negative aspects might be reduced by digitisation and its positive characteristics enhanced. As history shows, combat will always be intangible because of chaos and human fear. Developments in lethality, detection and precision in military technology will not remove these human elements.

Commanders are the focus for the human stresses of warfighting. Military leadership, with its imperative of personal example, means that a commander's behaviour symbolises to his subordinates the progress of a battle. Historically, the demands of field command have always been strenuous and demanding. In the future, the 24-hour battlefield will exacerbate the strain of command, because longer hours and a higher tempo of operations will inevitably eat more quickly into a commander's reserves of energy. The danger of fatigue impairing memory and mental performance will increase markedly.

Digitisation has potential to both limit and exacerbate the effects of stress on a commander. Digitisation promises to reduce the unknown in the field, and this will reduce a commander's degree of uncertainty and his fear of failure. However, such reduction of stress will depend largely on how successfully a culture of devolved command has been implemented. The potential of digitisation to exacerbate command stress will stem from the management of information. If operational information is not filtered, it will probably confuse rather than assist a commander's cognitive ability. Western armies have recognised the danger of digitisation producing information overload, which confronts them with the critical problem of how to focus digitisation to ensure that the process becomes the servant and not the master of both commanders and headquarters staff.

BIG-PICTURE BLINDNESS

The potential for commanders to narrow their focus below the level of their responsibility is real and dangerous. This is not simply a phenomenon that will affect commanders that are predisposed to micro-manage the actions of their subordinates. Under conditions of battle stress, where decisions may have potentially catastrophic consequences, most commanders will invariably focus on manageable or familiar areas of an operational problem. This tendency will mean that many senior field officers may fail to confront the macro level—or the full import and complexity— of a military operation.

In Western militaries, in which mastery of the subordinate level is a prerequisite for advancement, the likelihood of a commander descending to this area of operational management is real. A conservative solution would be to limit command devolution to subordinate operational detail, despite the capacity of technology to provide such detail. As an overall solution, imposing such a limitation seems attractive. Such a solution would focus the commander's attention on his own responsibility to master information and to execute faster decision making. The aim should be to ensure that the senior commander maintains the appropriate level of focus; yet this must be accomplished without denying the field commander the opportunity to exploit what might be a battle-winning opportunity by means of a 'bungy jump' descent to direct the activities of lower-level subordinates. There is a need to configure both battlefield combat information support (CIS) systems, with direction from an operational headquarters.

CIS systems must, in future, provide battlefield visualisation in a way that reinforces the individual commander's cognitive process. However, this process must take account of the realities of neuro-psychology. Configuration must serve the commander, but must retain sufficient flexibility for the system to be varied to each soldier's special capacities and strengths. Settings and access must continually reinforce an appropriate level of focus for each commander to maximise his performance in the field. A headquarters must be configured and responsibilities organised in order that the information from which the commander activates his OODA (observation, orientation, decision, action) cycle is presented in a coherent, efficient and timely manner.

Unfortunately, the natural inclination in armies when designing headquarters is to create filters whereby information overload of the commander is avoided. Historically, this process has always operated, but it risks limiting the potential of CIS systems to provide timely information. Selective filtering risks forfeiting the real advantage of digitisation. Indeed, filtering risks the very potential of the revolution in military affairs (RMA), which is to achieve disproportionate battlefield effectiveness through both greater operational tempo and more effective coordination of combat power. Operational tempo will not be enhanced if headquarters staff waste the advantage of information through the use of artificial filters without recourse to the traditional role of intelligence gathering.

During the RTA Trial there was a constant tension between the role of intelligence staffs and the potential of the all-informed battlefield. Intelligence staffs were rightly concerned that raw information might provide *situational awareness* but not necessarily *situational understanding*. There is a need for proper processing. During the trial, we were unable adequately to represent the enemy's offensive information operations (IO) threat and how his manipulation of information could affect our decision making. This problem revealed a fundamental dilemma for Western militaries that embrace digitisation—the need for continued operational risk. The factor of risk is one we all have great difficulty accepting, even in a peacetime military culture.

What is the solution to the problem of different processing times for information and intelligence? A resolution of this disparity is imperative in order to exploit tempo. During the RTA Trial it was found that one solution lies in acknowledging that both information and intelligence processes must coexist in parallel form. In short, information has to be passed in parallel to the intelligence cell and the commander and his headquarters staff. However, when displayed visually, information and intelligence needed to be separated by different colour codes, allowing the commander to appreciate the *degree of risk* emerging on the screen before him.

To minimise the risk of the commander acting intuitively on information rather than intelligence, the relationship between the commander and his intelligence officer (S2) was institutionalised through the creation of what was termed the 'Red Cell'. The relationship between the commander and the Red Cell became the most important function within the headquarters organisation. Consequently, the S2 became less a slave of tactical process in the headquarters and more of an intelligence adviser to the commander. The S2 physical proximity to the commander had the effect of creating a sounding board for the commander's decisions. The S2 provided an assessment of likely enemy responses based on available intelligence, while the commander retained an appropriate focus in his decision making. Overall, the relationship proved invaluable in helping the commander balance the risk of acting on information alone.

An unexpected consequence of digitisation, and the information technology (IT) command and control infrastructure that serves it, was the reduction in situational awareness through auditory means. Military headquarters have traditionally been distinguished by a babble of radio communications, all attempting to convey a battlefield situation. The urgency of the situation stems as much from the background battlefield noise and the emotion in the voices of the commanders or radio operators. Traditionally, all members of the headquarters, not just the radio operators, are in receipt of this information and are therefore situationally aware.

In the silent headquarters served by IT battlefield command and control systems, this traditional means of situational awareness is now largely lost. Those members not staring at a screen stand to lose a feel for the battlefield, and everyone is at risk of losing the import of particular situations on the ground. With this 'silent environment' comes the attendant danger that commanders and staff will begin to treat battle command and battlefield management as a computer game. This in turn risks creating an almost 1914–18 Western Front–style detachment between senior headquarters staff and front-line troops in terms of the realities and imperatives of war.

The phenomenon of the silent headquarters is the result of real limitations in equipment and bandwidth that mean that voice nets must be given up in order to provide digital nets. Eventually technology will overcome this limitation, but it will take time. Given that at present commanders and headquarters staff possess differing levels of familiarity with IT, it was decided that a voice command net was still necessary. The requirement for a traditional feel for the battle environment and the reality of the personal nature of command demanded that the voice command net be retained. While video display unit (VDU) links from brigade to battalion headquarters provided an even more powerful tool for imparting command intent, they do not replace the need for voice nets—until VDUs can be used by subordinate command vehicles in mobile situations. An ideal solution would be for the VDU to 'piggy-back' the voice net as an option for maximising commander-to-commander communication.

Other remedies for the loss of radios in the headquarters were more procedural. A multi-board was used at one end of the headquarters to constantly display battle maps and other selected information—including current intelligence priorities, overhead surveillance feeds and logistics bar graphs. Although these measures were very effective, the duty officer or operations officer continued to hold regular oral briefings of the whole headquarters staff while the latter remained at their workstations. This was not a new procedure. Many commanders use a similar device that is centred on their tactical 'bird table'. However, in a quiet headquarters oral briefings took on much greater significance. Indeed, they proved a powerful means of ensuring a common level of understanding of the operational situation and the commander's intent. The design of the headquarters was based on an open-plan room rather than the traditional 'T' or 'crucifix' shape. This was necessary because it was discovered that pockets of staff out of sight of the centre soon lost a capacity for situational awareness. Such isolation could not be tolerated in a headquarters where digitisation invariably meant a generation of greater tempo. It was critical to ensure that all staff had continuous situational awareness when on duty.

It was found that VDU links and an all-informed digital network offered opportunities to reduce the physical size of the headquarters. This reduction was accomplished by using what the US Marine Corps has called 'reach back'. During the RTA Phase 1 Trial, the legal officer, the chaplain, the senior medical officer, elements of the logistics staff and the plans section were all moved to a rear headquarters.[1] This measure was based on the belief that their advice could always be received through a 'reach back' process. This experiment was somewhat inconclusive. The chaplain, the senior legal officer and the legal officer jealously guarded their access to the commander; all saw access as essential to their respective tasks. However, success was achieved with the plans section operating in 'reach back'; it is possible that a cultural change may be required in future headquarters organisation.

The brigade's manoeuvre warfare focus meant that great emphasis was placed on forward and contingency planning. As a consequence, the commander's interaction with the plans section was continuous. Even the rudimentary VDU and collaborative planning tools available in the battlefield command system allowed the concept of 'reach back' to operate extremely well. A sizable part of the headquarters was relocated at a rear headquarters. The secure environment of a rear headquarters was much more conducive to work, demanding both white light and space. In future trials, the concept of 'reach back' should be pursued vigorously, employing all specialist personnel, including the analytical element of intelligence staffs.

There is no doubt that thoughtful application of digitisation techniques to the command relationship between brigade and battalion levels will enhance the coordinating role of commanders of these formations. Below battalion level, the leadership style must become increasingly participative. At company level, reaction time becomes shorter; the likelihood of communications between command vehicles manoeuvring for tactical advantage rather than to optimise communications diminishes. The impacts of combat and chaos are real and immediate. Nonetheless, even at platoon or troop level there is a requirement for a commander to maintain an overall understanding of his situation.

Phase 1 of the RTA Trial did not allow the consequences of digitisation to be tested at platoon or troop level. However, Phase 2 of the restructuring will incorporate these levels. Anecdotal evidence from US experiments suggests that there is every chance that troop commanders will spend more time inside the turret of a tank or armoured fighting vehicle examining screens than physically leading the fight against the enemy. A degree of digitisation must operate at company and troop levels. A company commander must, for instance, have reference to digital presentations to confirm his own and friendly forces' or enemy locations; he should also have the ability to pull down overhead visual feeds, giving pictures of his immediate front. But it must be recognised that, at company or troop level, commanders exercising personal observation of the fighting serve situational awareness best. Combat leaders will still fight tank, troop or squadron through vehicle optics or from personal observation of the unfolding action. What must be done in future is to ensure that digitisation is allowed to evolve to meet the needs of company or troop commanders, and not their superior headquarters.

SEARCHING FOR CERTAINTY

The second major effect of increased amounts of information available to a headquarters lies in the temptation for a commander to delay decisions in anticipation of acquiring the last piece of the operational puzzle. Of course, the puzzle can never be complete because of the fluidity of modern warfighting. A search for certainty would come to resemble the quest for the Holy Grail, and might negate the whole purpose of digitisation—which is to speed up operational decision making and tempo. As a recent British study noted, 'a future commander may spend so much of his time in the observing and orientating phases of the OODA loop that he either has less time to make adequate decisions, or is too late when he takes action'.[2]

The search for certainty is based on the expectation that digitisation will remove the effects of chaos. This notion is unrealistic and runs contrary to the nature of war in which Clausewitz's friction is omnipresent. The reality of incidental, random and unexpected effects in combat will inevitably degrade communications and reduce the effectiveness of digitisation. Throughout military history men have sought to introduce tactics, procedures and technologies to achieve more certainty in war. The use of radio is an obvious contemporary example; but even the Roman legion, the stockade and, later, medieval castles were, at root, attempts to exert order on warfare. As technique proved its worth, it spawned countermeasures and imitation, which eventually reduced its relative advantage.

The military art has essentially evolved despite the eternal problems of chaos and uncertainty. Fundamental military principles and practices such as reserves, mass, main effort and mission control all acknowledge the chaotic nature of war, and aim to win despite the uncertainty of human combat. We must not lose the ability to fall back on such tried methods when digitisation either fails or is prevented from delivering superior situational awareness. At the same time, it is vital that we avoid creating a generation of commanders that place infallible faith in the digital icon to fill information gaps. The greatest strength of the military profession is its recognition that there will always be gaps in knowledge; tactics and procedures exist to overcome uncertain conditions.

In war, digitisation remains only a relative advantage. Historically, commanders have been tested on how they have applied tactical principles. Paradoxically, the potential for a more informed battlefield may also mean greater accountability in the wake of battle. A board of inquiry can now audit a commander's decision against a great store of information. Yet at the time of decision, the commander may have been uncertain of the data's veracity. While his historical predecessors possessed scant information, a future commander may suffer from a surfeit of information. We should not underestimate the additional stress this will place on decisions and how commanders are likely to react as a result.

In the future, both military training and culture must adapt to the problem of an information-rich but not necessarily clearer operational environment. Military training must provide commanders with real tools and skills for improved risk management. Future training and education must emphasise maintenance of the aim, and develop the skill and determination to 'screen out' issues peripheral to focused decision making. Military culture must continue to emphasise the need to make decisions without complete information, even apparently essential information. The modern military must not succumb to the ideal of a science of warfare. We must retain the art of warfare along with the strategy, tactics, intuitive skills and operational procedures necessary to achieve victories. The need to retain warfare as an art will probably necessitate improved selection procedures for potential combat leaders. Candidates must possess the qualifications necessary for field command.

Military culture must emphasise risk management as a normal function of the profession of arms. Yet this must be done in such a way as to avoid creating a risk-adverse military establishment, which can be difficult to achieve in peacetime militaries. In the future, if we intend to train commanders that can apply the art of war to decision making in situations of imperfect knowledge, we are looking for creative individuals who can interpret patterns and distinguish the important from the ephemeral. In such a process, digitisation offers assistance to future commanders through visual presentations that illustrate developing patterns and emerging situations on the battlefield.

CONCLUSION

Digitisation of the battlefield will enhance the quality of combat decision making. However, to achieve the full potential of digitisation, modern armies must accept that the realities of the battlefield and the function of battlefield leadership are unlikely to change radically. Digitisation must be focused and directed for maximum effect at each level of command. The process must not be allowed to become another stovepipe system that exists only to serve the information requirements of higher headquarters.

When analysing the phenomenon of digitisation, it is difficult not to see the ghost of Jomini arising from the ashes of military history. It was Jomini that placed faith and hope in the ascendancy of the tangibles of the science of war—as opposed to Clausewitz, who emphasised the intangible elements surrounding war as an art. The RTA Trials, and I believe overseas experience in similar endeavours, reveal that the Jominian science of war is unlikely to prevail in all but the most unequal of military contests. Even in the information age, battle remains the supreme contest of human wills, with its clash between measure and countermeasure ensuring that chaos will endure. Advanced militaries must, at all costs, preserve and develop their ability to operate and to triumph in that chaos.

9 | Stress on higher commanders in future warfare

David Horner

IN NOVEMBER 1965 the US Joint Chiefs of Staff were granted a private meeting with President Lyndon Johnson. They were extremely concerned that US policy in Vietnam was leading to a slow build-up of forces in a war of attrition. Rather than allowing a protracted Asian ground war to develop, they believed that, if the war was to be won, the USA had to use its air and naval power to punish the North Vietnamese.

By chance, there was a witness to this private confrontation between Johnson and his military advisers. He was Major Charles Cooper, ADC to the US Chief of Naval Operations, who was there merely to carry a large briefing map, which he was to place on an easel before vanishing from the scene. But when the easel could not be found, Johnson ordered Cooper to hold the map; as Cooper recalled, 'I had become the human easel—one with ears'. Cooper later became a Marine Corps lieutenant general and, some 30 years later, recalled how the Chairman of the Joint Chiefs of Staff, General Earle Wheeler, put his case. The other chiefs, including the US Army Chief of Staff, General Harold Johnson, supported Wheeler. As soon as the chiefs had finished, President Johnson 'whirled to face them and exploded':

> I almost dropped the map. He screamed obscenities, he cursed them personally, he ridiculed them for coming to his office with

134

their 'military advice'. Noting that it was he who was carrying the weight of the free world on his shoulders, he called them filthy names—sh-heads, dumbsh-s, pompous assh-s—and used 'the F-word' as an adjective more freely than a Marine at boot camp. He then accused them of trying to pass the buck for World War III to him. It was unnerving. It was degrading.

The interview continued with more exchanges and abuse from Johnson, before he ordered the chiefs to 'get the hell out of my office'. The policy in Vietnam remained unchanged. As the Chief of Naval Operations drove back to the Pentagon he told Cooper that he had known tough days in his life and sad ones, but 'this day has got to be the worst experience I can ever imagine'.[1] The US Army Chief, Harold Johnson, was similarly affected. In retirement he said:

> I remember the day I was ready to go over to the Oval Office and give my four stars to the President and tell him, 'You have refused to tell the country they cannot fight a war without mobilization; you have required me to send men into battle with little hope of their ultimate victory; and you have forced us in the military to violate every one of the principles of war in Vietnam. Therefore, I resign and will hold a press conference after I walk out of the door' . . . I made the typical mistake of believing I could do more for the country and the Army if I stayed in than if I got out. I am now going to my grave with that lapse in moral courage on my back.[2]

Clearly, the sort of stress faced by higher commanders differs from that endured by battlefield commanders. Unlike battlefield commanders they generally sleep in a comfortable bed at night, eat three good meals per day and are rarely in danger from the enemy. Most times, they are not woken at night to be told that the enemy has broken into the forward positions; they are not thrown into despair when the planned resupply is interrupted by bad weather; they do not see men with shattered limbs being carried past; their command post and they do not write letters of condolence to families of men who they have served with personally in difficult times. To be fair, higher commanders have almost certainly had their apprenticeship as tactical commanders. The point is that, as higher commanders, they do not escape stress.

DEFINITIONS

Before beginning as discussion on stress on higher commanders in future warfare, we need to be clear on the meaning of two key terms: stress and higher command. Stress has a number of meanings. It includes the force that is exerted on a body (i.e. it is an outside influence that disturbs the natural equilibrium of the body). But stress is also the body's response to that outside force. In general terms, then, stress is the relationship between an environmental stimulus and the individual's reaction to it; or, as one psychologist puts it, 'stress is part of a complex and dynamic system of transactions between the person and his environment'.[3] In the psychologist's jargon, the environmental factors are the stressors that cause the individual to suffer stress. Of course, different people react to stress in different ways. Here I am not going to try to take up the psychologist's role of assessing the personality factors that contribute to different capacities to handle stress. Nor am I going to try to describe the physical and psychological manifestations of stress. It is sufficient to recognise that, while some level of stress is useful in achieving an optimum performance, excessive stress certainly degrades performance, and in higher commanders this might have disastrous consequences.

This chapter concentrates on the stressors that affect an individual—in this case a commander—in the workplace. In general terms these fall into the following categories.[4] First, they relate to working conditions, and in particular to work overload. This might include quantitative overload (i.e. too much to do); and qualitative overload (i.e. work that is too difficult). The second category concerns the individual's role in the organisation. Stress is increased if the individual's role is ambiguous (i.e. if there is a lack of clarity in work objectives). Stress is also increased if there are conflicting job demands, or if different superiors make different demands. Responsibility brings its own stresses, and these have been found to be greater when the individual is responsible for personnel than when responsible for things. This is particularly important for commanders that are responsible for soldiers' lives.

The third category concerns work relationships with superiors, subordinates and colleagues: these relationships seem to increase stress as an officer rises in rank. Career development problems can also be a cause of stress: for example, lack of job security or frustration at reaching career ceilings—a well-known military phenomenon. An organisation's structure can be a cause of stress: for example, if an individual feels a powerless cog in a giant machine. Finally, an individual is subject to outside stress such as family and personal problems, even if these do not relate directly to his work. This chapter considers these stressors by describing them in practical terms as problems faced by higher commanders.

The other term that needs explanation is higher command. This chapter concentrates on the operational level of command, but also draws freely on the experiences of strategic-level commanders where necessary, as I have done with my opening story. Also, it is appropriate to interpret the operational level fairly broadly. It includes what the Americans call the theatre strategic level of command. Some examples at this level would include the commanders-in-chief in World War II, such as Generals Douglas MacArthur and Dwight Eisenhower, as well as General Norman Schwarzkopf in the 1991 Gulf War.

Below this level is the operational level in its pure form—that is, the command of campaigns. Generals Montgomery and Patton are classic examples. Australia has had few operational-level commanders with responsibility for the planning and conduct of campaigns. Generals Blamey, Rowell and Herring all conducted campaigns in New Guinea in World War II but, in each case, operated in an Allied command structure headed by MacArthur.

While few Australians have actually commanded campaigns, many others have served as national commanders. In some circumstances a national commander might occupy purely an administrative position. However, in other circumstances he might have to interpret the strategic mission of his force—as determined by the government—in the light of the strategic and operational objectives of a theatre commander from an allied nation. In this sense, the national commander operates *at* the operational level, even if he is not *the* actual operational-level commander. Thus, as GOC of the Second Australian Imperial Force (AIF) in the Middle East, General Sir Thomas Blamey was the national commander and, in his negotiations with the British Commander-in-Chief, General Wavell, was at the operational level. Similarly, the Commander Australian Force Vietnam was at the operational level even though he had no clear operational-command responsibility.

Higher command in war has always been stressful. It is for this reason that Field Marshal Wavell, in setting out the qualities of a good commander, thought that the first essential was 'the quality of robustness, the ability to stand the shocks of war'. As he put it, 'the general is dealing with men's lives, and must have a certain mental robustness to stand the strain of this responsibility'.[5] Field Marshal Montgomery called it 'toughness',[6] and Napoleon had the same idea:

> The first qualification of a general-in-chief is to possess a cool head, so that things may appear to him in their true proportion and as they really are. He should not suffer himself to be unduly affected by good or bad news.[7]

Field Marshal Slim thought that a higher commander needed the qualities of willpower, judgement, flexibility of mind, knowledge and integrity. In explaining integrity he said that to maintain the confidence of your men you needed integrity: 'All the really great commanders who have held their men have had it because the only foundations under men which will stand under great stress are the moral ones'.[8] Major General J.F.C. Fuller, in his wonderful little book *Generalship, Its Diseases and Their Cure*, saw three pillars of generalship: courage, creative intelligence and physical fitness, and observed that they were the attributes of youth rather than of middle age. Fuller took the ages of a hundred generals, from Xenophon (401 BC) to Moltke (1866), and discovered that on the average the period of most efficient generalship lay between the years of 30 and 49, and that the peak was reached between the years of 35 and 45.[9] Fuller was working on the theory that, within reason, a younger commander is more likely to be able to handle a stressful situation than an older one.

FACTORS THAT CONTRIBUTE TO STRESS

It has become a cliché to say that predicting the future is a risky business. I do not believe that I am smart enough to make even a fair attempt at it. Rather, my approach is to begin by looking at what has happened in the past, in the belief that an understanding of how we have arrived at our present position is the first step towards looking at the way ahead. I begin, therefore, by trying to discern the environmental factors or stressors that caused stress to higher commanders in the past. I then consider how or whether these environmental factors are changing, as a means of speculating about stress on higher commanders in future war. There are at least 10 factors that contribute to that stress, although many of them overlap.

The enemy

The enemy factor might seem self-evident: after all, if there were no enemy there would be no war. But the enemy does not usually strike directly at the higher commander. Sometimes the enemy tries to assassinate the higher commander, but this is rare and success is even rarer. A classic example is the failed commando raid to kill the famous German general Erwin Rommel in November 1941. One success was the aerial ambush of Admiral Yamamoto in 1943, which was possible only because the Americans had broken the Japanese-enciphered radio traffic. An enemy will probably conduct bombing attacks on the higher commander's headquarters (if it can be located), but measures can be taken, and usually are, to reduce this threat.

Stress on higher commanders is brought about not by direct enemy attack on them but by the consequences of enemy action. As Fuller put it,

> the enemy does not attack [the commander] physically, but mentally; for the enemy attacks his ideas, his reason, his plan. The physical pressure directed against his men reacts on him through compelling him to change his plan, and changes in his plan react on his men by creating a mental confusion which weakens their morale.[10]

It was for good reason that General André Beaufre described strategy as 'the art of the dialectic of two opposing wills using force to resolve their dispute'.[11]

Stress comes from finding that the enemy has moved faster, or in an unexpected way. To use the modern jargon, stress comes from finding that the enemy has moved within the commander's Boyd decision–action cycle, also known as the observation–orientation–decision–action (OODA) loop.[12] Part of the stress placed on General MacArthur and his staff at GHQ Southwest Pacific Area in August 1942 was the realisation that the Japanese were halfway over the Owen Stanley Range before they had taken the threat seriously. Earlier, GHQ had obtained intercepted messages warning that the Japanese were contemplating this approach, but the warning had become lost in the 'noise' of other information.

High commanders are also placed under stress when an enemy reacts in an unconventional manner. Part of the stress on General Westmoreland in Vietnam was imposed by his inability to pin the Viet Cong down in set-piece battles.

Resources

A commander's ability to conduct his campaign is partly determined by the resources allocated to him. In an ideal world a
commander should be given sufficient resources to conduct his
campaign, but that will not usually be the case. Thus, MacArthur
had to defend New Guinea, but had insufficient maritime resources
to reinforce New Guinea and to maintain large numbers of
formations once they arrived there. Rommel had to remain on the
defensive at El Alamein when he ran out of fuel. Similarly, General
George Patton had to curtail his 3rd US Army's advance in northern France in September 1944 through insufficient fuel.

These resource limitations require the commander to make
tough and stressful decisions about priorities. A key resource is
personnel—or, as it used to be known, manpower. Commanders
have always been concerned about maintaining force strengths.
Some of General Blamey's biggest arguments with the government
were over manpower.

Closely linked to personnel strengths—and one subject to highly
stressful decisions—is the level of acceptable casualties. The commander has to weigh the loss of lives against possible alternatives.
He has to consider not just the human cost but whether he has
sufficient reserves to make good the casualties, and what level of
casualties would be acceptable. The Commander Australian Force
Vietnam was not given any guidelines about acceptable casualties
from the government or the Australian chiefs of staff. Nonetheless,
in July 1970 the then Australian commander, Major-General
C.A.E. Fraser, expressed reservations to the US headquarters in
Vietnam about proposed operations by the Australian taskforce
in the Long Hai hills, noting that past operations there had 'been
costly in life and productive of limited military gains'. There were
no more operations in that area.[13]

An important subset of the resource factor is the level of
technology available to the commander's forces. The greatest stress
faced by higher-level commanders on the Western Front in World
War I was in devising ways of overcoming the stalemate of trench
warfare. This stalemate related to a mismatch in technology. New
technology such as the machine gun, artillery and barbed wire
made it extremely difficult for an attacker to break an enemy's
defensive line. It was several years before techniques could
be developed to make better use of artillery and before other

technology such as the tank and aircraft could be used to break the stalemate.

A critical technology is that of communications. While the fog of war will always exist, it is made worse if the commander has difficulty communicating with his superior headquarters and subordinate formations. Part of the pressure faced by General Rowell in August 1942 was the lack of up-to-date information from his formation commanders fighting at Milne Bay and on the Kokoda Trail. By the 1991 Gulf War the commander of the Australian naval task group could communicate directly and often with the Australian Maritime Commander in Sydney. Whether that eased or exacerbated his stress is arguable.

Physical limitations

The physical limitations on a commander's ability to conduct his campaign include the climate, terrain and extent of the area of operations. These factors complicate the planning and conduct of the campaign. For example, the physical environment of New Guinea led to heavy losses from malaria and other tropical diseases during the Papuan campaign in 1942–43, and placed the operational commanders under greater stress as they sought to maximise their resources and bring them to bear on the enemy.

Based in Brisbane, it was not possible for MacArthur to visit the forward troops on the Kokoda Trail and hence to appreciate their difficulties. Blamey was GOC of the AIF in the Middle East for 20 months, from mid-1940 to early 1942. During that time he covered over 77 000 kilometres by air, over 32 000 kilometres by car and 7500 kilometres by sea to inspect and command his forces.[14] The mere physical effort of commanding his forces was immense. During the campaign in New Guinea Blamey tried to relieve senior commanders every four months because he believed that that was as long as they could operate effectively in the trying climate. In Vietnam the Australian commander in Saigon could visit the headquarters of the Australian taskforce in the field by helicopter after breakfast and be back in Saigon for lunch.

So far we have mentioned several fairly predictable factors which, except for the degree of their importance, could relate as much to tactical- as to operational- or strategic-level commanders. We now come to factors that relate more to the higher levels of command.

Political pressure

In previous centuries operational-level commanders were largely immune from this pressure on a day-to-day basis. British commanders in India or General Arthur Wellesley in the Peninsular campaign could conduct their operations knowing that it could be several weeks before contrary directions, queries or criticism might arrive by ship from London. By the time of the American Civil War, however, the telegraph enabled the US president to keep in close touch with his senior commanders. President Lincoln has been criticised for abusing this new capability, but undoubtedly by doing so he saved the Union. Nonetheless, he understood completely the different roles of politicians and generals. His views are set out in a famous letter to General Ulysses S. Grant:

> The particulars of your plans I neither know nor seek to know. You are vigilant and self-reliant; and, pleased with this, I wish not to obtrude any constraints or restraints upon you. While I am very anxious that any great disaster, or the capture of our men in great numbers shall be avoided, I know these points are less likely to escape your attention than they would be mine. If there is anything wanting which is within my power to give, do not fail to let me know it. [15]

Inevitably, this idealised model has rarely been followed. The bitter controversy between the British Prime Minister Lloyd George and his Chief of the Imperial General Staff, Field Marshal Sir William Robertson, in World War I is a prime example of the difficulties of modern civil military relations. In World War II, Winston Churchill dealt directly with the commanders-in-chief in the field. He constantly harassed General Wavell in the Middle East with directions and queries. After the British were forced out of British Somalia by the Italians in August 1940 Churchill sent a 'red-hot cable' to Wavell, complaining that the lack of casualties showed that his forces had not fought well. Wavell replied testily that 'a big butcher's bill was not necessarily evidence of good tactics'. The Chief of the Imperial General Staff said that this reply roused Churchill 'to greater anger than he had ever seen him in before'. [16] Wavell had that robustness that he lauded in his lectures on generalship, but certainly felt the pressure of Churchill's constant harassment.

Closer to home, the effect of political pressure on commanders was demonstrated during the Owen Stanley campaign in 1942. As the Japanese advanced over the Kokoda Trail towards Port Moresby, the Australian politicians became increasingly nervous. Billy Hughes, the former prime minister and leader of the opposition United Australia Party, stated publicly that there had 'been a lamentable lack of vision, of initiative, of coordination of control by our military leaders. They have failed to anticipate the enemy's movement'.[17] Concerned at the turn of events, the Advisory War Council (with members of both government and opposition) summoned the Australian Army Commander-in-Chief, General Blamey, to Canberra to report on the operations. Blamey was directed to go to New Guinea to gain a first-hand account of the operations. On his return he again reported to the Advisory War Council. By this time the operational situation had deteriorated further and Blamey received a hostile reception from various politicians. After he left, one senior minister said, 'Moresby's going to fall. Send Blamey up there and let him fall with it'.[18] A few days later Frank Forde, deputy Prime Minister and Army Minister, told a senior army officer that if the Japanese took Port Moresby it would be a serious situation: 'I'd lose my seat in Capricornia'.[19] As we know, Prime Minister John Curtin, at the suggestion of General MacArthur, ordered Blamey to New Guinea to take command there, which led to Blamey's dismissal of the operational commander, General Rowell. Before leaving for New Guinea Blamey wrote to Rowell that his arrival did not indicate a lack of confidence in him. 'It arises', wrote Blamey, 'out of the fact that we have very inexperienced politicians who are inclined to panic on every possible occasion'.[20]

Blamey was never free from political pressure. At times the Army Minister, Forde, demanded special considerations for his political constituents, who wanted sons or husbands released from the Army or given particular postings. Later in the war Blamey was criticised by opposition members in parliament as a means of attacking the government. As a strategic-level commander Blamey could not escape being faced with constant political pressure. At least he was able to prevent the politicians from applying their pressure to his subordinate commanders. In fact very few Australian operational-level commanders have had to deal with direct political pressure. But political pressure has often been applied indirectly through the strategic-level commanders to the operational commanders and even to the tactical level. When, in December 1942, a battalion commander was preparing to attack Gona and was fearful of heavy and unnecessary casualties, his divisional commander, Major-General George Vasey, sent him a short note:

'Canberra must have news of a clean up and have it quick or we will both go by the boot'.[21]

The Papuan campaign gives some idea of the nature of the stress engendered when senior commanders transferred their concerns to subordinate operational commanders. During the Milne Bay battle in late August 1942, General Rowell, the commander of New Guinea Force in Port Moresby, received orders from Brisbane that he was to take the offensive. Rowell replied privately to Blamey's deputy in Brisbane: 'I'm personally very bitter over the criticism from a distance and I think it damned unfair to pillory any commander without any knowledge of the conditions . . . I suppose there will be heresy hunts and bowler hats soon'.[22] A month later, when Rowell was relieved of his command, he wrote, 'I have had considerable heart searching during the past 48 hours as to whether I should have gone on eating dirt . . . I can hardly accept a position where my self respect is lost'.[23]

Politicians apply pressure to military commanders for a number of reasons. They might be trying to score points as part of an effort to discredit the government or a particular minister. (This was behind the attack on Blamey in parliament in early 1945.) They might be driven by the need to win an election. A military success is handy for an incumbent government just before an election. They might be trying to satisfy discontented constituents. But the most important reason is the need to respond to public opinion. In turn this relates to how much information is available to the public.

During the 20th century the public progressively received more information about military operations and at a faster rate. The landing by Australian troops at Anzac Cove on 25 April 1915 was announced publicly in Australia on 29 April, although the first lengthy accounts did not appear until 8 and 9 May.[24] Ultimately, however, it was an Australian newspaperman, Keith Murdoch, who reported on events at Gallipoli to the British Prime Minister and who contributed to the replacement of the British Commander-in-Chief, General Sir Ian Hamilton. During the 1991 Gulf War the US television company, CNN, was broadcasting military developments faster than they could be relayed by the military chain of command.

Control of the media

While the increasing effectiveness of the media has led to politicians applying more pressure to commanders, the need to manage and control the media itself has also contributed to stress on higher commanders. Commanders know the power of the media to affect public opinion and potentially to release information that might prejudice the security of their operations. Initially, commanders dealt with this stress either by banning the media or by strictly controlling it. General MacArthur insisted that nothing could be released unless it had first been announced in his communiqués, which he wrote himself. Both MacArthur and the Australian Army employed public relations staff. By contrast, senior Australian commanders in World War I did not employ special public-relations or public-information officers. By the time of Australia's commitment of a battalion group in Somalia in early 1993, the highest priority was given to the deployment of a Media Support Detachment. As Bob Breen, the Army historian of the operation, wrote, '[Defence Minister] Senator Robert Ray, possibly with an eye on the forthcoming election, ordered General Gration [the CDF] to coordinate maximum media coverage' for the operation.[25] Yet the Commander of the Australian Force, Colonel Bill Mellor, had no command authority over the Media Support Detachment.[26] Control of the media is now one of the first concerns of any commander planning a military operation. It is particularly stressful when commanders realise that they cannot control the media.

Public support

Although an operational commander should be able to conduct his campaign free of non-military concerns, in practice this is not possible. Increasingly, commanders have had to consider the issue of public support, even though theoretically this is a matter for politicians. This issue was not quite so important during the world wars, when there was almost total public support. Maintaining public support has become more critical as forces have been deployed for political purposes rather than for the immediate defence of the homeland. The issue goes beyond the role of the media: a commander needs to be aware that operations carried out by his forces might affect the public support for his forces. The destruction of the Viet Cong–dominated village of Long Phuoc in June 1966, however justified it might have been in military terms, always carried the danger of decreasing Australian public support for the war.[27]

Heavy casualties, especially if incurred for no apparent gain, will also affect public support. The effect of American casualties in Somalia is a case in point.

Cooperation with allies

Working with allies has always been a source of stress and tension. Wellington found it to be so when working with the Portuguese in the Peninsular campaign. The disagreements between Eisenhower and Montgomery have gone into legend. Field Marshal Slim, a much easier person to deal with than Montgomery, summed it up this way:

> it is an extraordinary thing that you should meet with so much opposition from allies. Allies, altogether, are really very extraordinary people. It is astonishing how obstinate they are, how parochially minded, how ridiculously sensitive to prestige and how wrapped up in obsolete political ideas. It is equally astonishing how they fail to see how broad-minded you are, how clear your picture is, how up-to-date you are and how cooperative and big-hearted you are . . . But let me tell you . . . that you are an ally too, and all allies look just the same. If you walk to the other side of the table, you will look just like that to the fellow sitting opposite.[28]

Working as a junior partner in a coalition brings additional stress. Blamey spent a vast amount of his time in the Middle East dealing with problems concerning his relations and those of his force with the British. His exchange with General Auchinleck at a conference about the relief of Australian troops at Tobruk in September 1941 is well known. Auchinleck began by stating that Tobruk could not be relieved. 'Gentlemen', said Blamey, 'I think you don't understand the position. If I were a French or an American commander making this demand what would you say about it?' 'But you're not', replied Auchinleck. 'That is where you are wrong', said Blamey. 'Australia is an independent nation. She came into the war under certain definite agreements. Now, gentlemen, in the name of my Government, I demand the relief of these troops.' Auchinleck shrugged and said: 'Well, if that's the way you put it, we have no alternative'.[29]

In New Guinea Blamey had to deal with the Americans, and initially suffered in silence as MacArthur criticised his forces withdrawing on the Kokoda Trail. After the Americans failed in their first battle Blamey saw his opportunity. When MacArthur wanted to send in more Americans, Blamey told him bluntly that 'he would rather put in more Australians, as he knew they would fight'.[30]

Host-country authorities

When forces are deployed overseas, operational and national commanders are also likely to spend much time dealing with
host-country authorities. If it is assumed that the forces are
deployed at the request of the host country, the causes of tension
might be minimal; but even then there might be tension caused
by differences of interpretation of the mission, and problems caused
by the influx of large numbers of troops with different economic
standards and cultural norms. In his autobiography, General
Schwarzkopf recalled that when he and his forces arrived in Saudi
Arabia he had to mask his sense of urgency in his dealing with
the Saudis. 'To my consternation', he wrote, 'their most pressing
concern was neither the threat from Saddam nor the enormous
joint military enterprise on which we were embarked. What
loomed largest for them was the cultural crisis triggered by the
sudden flood of Americans into their kingdom'.[31] Failure to deal
adequately with this issue had the potential to destroy the whole
purpose of the deployment.

Under the same heading might be included the problems of
dealing with local civilian authorities when forces are deployed in
their own country. For instance, the commander of Australia's
Northern Command has to deal with the Northern Territory
government and local Western Australian authorities in the north
of that state. In World War II this problem was overcome by
placing the civil administration in the threatened area under military control, but even then the task became a burden for the
military commander.

Military jealousy and pride

At first glance the inclusion of jealousy and pride as causes of stress
on higher commanders might seem absurd. After all, when lives
are at stake, let alone when the security of the country is at risk,
it might seem difficult to believe that commanders would allow
their personal concerns to influence their decisions and cause undue
stress to them and their subordinates. Nothing could be further
from the truth. Higher commanders reach their positions because,
in the main, they have strong egos and self-belief.

The arguments between Montgomery, Eisenhower, Tedder,
Bradley and Patton during the north-west Europe campaign in
1944–45 related as much to personal foibles as they did to differences between allies and honest disagreements over the most

effective way of defeating the Germans. Australian military history is replete with similar instances. In the Middle East in March 1941 Blamey selected the 6th Division to go to Greece because, among other things, he did not want to give an operational opportunity to the commander of the 7th Division, Major-General John Lavarack. Blamey then tried to appoint Lavarack as GOC of the AIF Rear Headquarters. Fortunately, Blamey gave Lavarack the opportunity of refusing. 'I refused at once', wrote Lavarack in his diary. 'What sort of person does he think I am?'.[32]

The problem is that many commanders actually believe that they need to remain in command for the national good. During the Papuan campaign, Blamey believed that he had to relieve Rowell, Major-General Allen and Brigadier Potts so as to maintain his position as Australian commander-in-chief. Blamey was under threat from MacArthur, who also believed that his position was under threat. MacArthur feared that another reversal might result in his dismissal, just as Admiral Ghormley, the commander of the neighbouring South Pacific Command, had been dismissed. A few months later, MacArthur demanded quick results at Buna, Gona and Sanananda so that he could achieve a victory in his theatre before the US Navy secured theirs at Guadalcanal. Not only was MacArthur under stress, but he transferred it through Blamey to the Australian senior officers and directly to his own subordinates. His orders to his US corps commander, Lieutenant-General Robert Eichelberger, are well known: 'I'm putting you in command at Buna . . . I want you to take Buna, or not come back alive'.[33] Eichelberger then had the additional stress of dismissing his West Point classmate, Major-General Harding.

The importance of the campaign

The last factor that places stress on a higher commander is the importance of the campaign. The senior commanders involved in the Greek campaign, Generals Wavell, Wilson and Blamey, were distressed that it turned into an ignominious withdrawal and evacuation, with many fine troops being captured. They knew, however, that despite the defeat the British Empire would be able to continue the war. Within days they were able to recover from the defeat and focus on new campaigns.

It was a different matter when Rowell faced the prospect of the Japanese seizing Port Moresby, from which they could have bombed northern Australia. It was different again in Vietnam.

Whatever happened in Phuoc Tuy Province, Australia's security was not directly at stake.

Summary

In listing these 10 factors that contribute to the stress on higher commanders, I am conscious that there might be others and also that no single listing of factors adequately sums up the particular stress of commanding large forces in war. The classic example was the stress placed on General Eisenhower as he pondered whether to order the Normandy landing to take place on 6 June 1944, when a mere glance out of the window showed a nasty storm that, if it persisted, would wreck the landing with calamitous results.

Eisenhower's experience should be compared with General Schwarzkopf's in the Gulf War. In his autobiography, Schwarzkopf recalled a private telephone call to the Chairman of the Joint Chiefs of Staff, General Colin Powell, a few days before the ground attack:

> The minute I began with, 'We're having a problem with the weather', Powell became exercised.
>
> 'I've already told the President the twenty-fourth. How am I supposed to go back now and tell him the twenty-sixth? You don't appreciate the pressure I'm under. I've got a whole bunch of people here looking at this Russian proposal and they're all upset. My President wants to get on with this thing' . . .
>
> I got pretty exercised too. 'I'm not trying to be a smart-ass, but what if we attack on the twenty-fourth and the Iraqis counterattack and we take a lot of casualties because we don't have adequate air support? And you're telling me that for political reasons you don't want to go in and tell the President he shouldn't do something that's militarily unsound?' . . .
>
> 'Don't patronize me with your talk about human lives!' he shouted. It was the first time I'd ever heard him lose his temper, and he was livid. 'What are you doing? Sitting there in front of all your officers putting on a bit of a show while you talk to me this way!'
>
> I got hot too . . . 'I'm not doing that at all and I'm not being disloyal to you. What I'm trying to say is that I'm under pressure too' . . .
>
> 'Sometimes I feel like I'm in a vise—like my head is being squeezed in a vise. Maybe I'm losing it. Maybe I'm losing my objectivity. But I don't think so.'

By this time Powell had calmed down. 'No, no, you're not losing it', he told me. 'You haven't lost anything. I have great confidence in you.'

Half an hour later the weather forecasters changed their minds, and Schwarzkopf called Powell back to tell him that the attack would go ahead on the 24th.[34]

I think that exchange gives a good idea of the nature of stress faced by modern operational-level commanders. I am sure that Rowell in Port Moresby would have liked to have had a similar telephone conversation with Blamey in Canberra in 1942, rather than rely on cabled messages.

One commentator compared Eisenhower and Schwarzkopf:

> The modern general has state-of-the-art technology light years ahead of what Eisenhower had available to him, yet both faced the same questions—*Is it time? Has enough been done to ensure success of the attack?* Both had huge staffs manned with the best minds they could assemble; both sought and used the best intelligence they had available to them, but the decision to commit forces to the offensive in the end was largely intuitive, personal and private—*In my judgement the time is right.*[35]

THE FUTURE

Having considered the factors that contribute to stress on higher commanders, we should consider how those factors are changing or are likely to change in the near future.

The enemy—the future

The first factor, you will recall, was the enemy. Some analysts consider that the Gulf War was the last of the large-scale conventional wars. Whether that is so or not, there is unquestionably a greater likelihood of asymmetrical wars or threats.[36] An enemy will be less likely to present himself to attack in the open. Indeed, it might even be hard to classify the enemy as such. The bandits faced by the 1st Battalion of the Royal Australian Regiment in Somalia are an example of such a non-traditional threat. There is also the possibility that the threat might come in cyberspace, from an enemy that cannot be identified. Dealing with such threats will present special problems for future commanders. They will be required to think beyond the neat problems presented in staff colleges.

Linked to the obscure status of some potential opponents is the problem presented by ambiguous missions. Mission creep or mission drift is a phenomenon that occurs when directing authorities progressively change and expand a force's mission. It can impose considerable pressure on the commander if the new objectives are significantly outside the original mission and require unjustifiable resources or risks.

Resources—the future

The second factor was resources, including personnel and technology. Maintaining personnel levels is likely to be an even more difficult problem than in the past. Personnel problems have become more complex, with changing norms and the increasing role played by women in society.

As Hugh Smith describes in Chapter 4, Western governments and societies are now far less tolerant of casualties than in earlier times. Some might argue that casualties would still be acceptable if the cause were important enough. It is unlikely that Australian forces will be fighting clearly for the defence of Australia, and likely that they will be deployed for political purposes which, almost by definition, will be subject to argument and ambiguity. In these circumstances the loss of lives is even more difficult to justify. At the same time the experience of the Gulf War and the development of precision-guided munitions (not to mention the extravagant claims of some airpower advocates) have created a climate of expectation that there will be few casualties. The stress of being responsible for the loss of lives will continue.

In recent years, financial accountability has become more important for military commanders and will impose additional stress on them. Before the Somalia operation the Australian Defence Department was asked for an estimate of the operation's cost. This was put at $20 million, and therefore anything above this figure had to be met by the Australian Defence Force's current budget allocations. Bob Breen, in his account of the Somalia operation, describes how Air Headquarters was reluctant to move stores to Somalia by air because it had not been informed that $2.4 million had been allocated to it to pay for airfreight.[37] By contrast, General Rowell in New Guinea in 1942 would probably never have thought about the cost of conducting his campaign.

Technology—the future

The introduction of new technology has always placed strain on higher commanders. But the development of information technology has had a profound effect on command arrangements and on commanders, to the extent that we must consider it as a separate factor. Communications have now improved to a point where individual soldiers in an infantry platoon can speak to each other by radio. It is also possible for a prime minister or president to speak directly to platoon commanders in battle. In response to these developments, defence organisations in Western countries have put in place command structures that are specifically designed to limit the capacity of politicians and strategic-level commanders to interfere at the tactical level. This worked satisfactorily for both the Americans and the Australians in the Gulf War. If anything, the command structures have been tightened even more. For instance, the establishment of Headquarters Australian Theatre, located outside Canberra, has underlined the role of the operational-level commander. Whether these neat arrangements would persist in time of major and politically sensitive operations is open to doubt. Higher commanders would then face the prospect of being bypassed in the chain of command, yet still being held responsible for the activities of their command.

In Chapter 8 Brigadier Wallace discusses the effect of digitisation on commanders, and I do not intend discussing it in detail. It is probably a reasonable assumption that commanders in the future will have a high level of understanding and competence in the use of digitised displays and other command and control technologies. One US military psychologist has speculated that in the future the army might have superb battle staff officers, who will 'know how to leverage powerful analytical tools for tremendous advantages in speed, precision and effect'. He suggests, however, that this transformation in the capabilities of officers might cause problems when they move to higher levels of command, where the key issues relate not to carefully displayed information but to issues requiring intuition and the factoring in of political and other information that does not lend itself to computer display. As he puts it, the army might 'produce a generation of senior leaders that is so insecure without their computer models and decision systems that they could not step beyond them'. Future leaders, brought up under 'the illusion of omniscience from multi-sensory information systems', might 'fear the "guesstimate", preferring to avoid risking mistakes by substituting certainty models for their intuition'.[38]

New technology has also increased the amount of information available to commanders. There is certainly a danger that commanders might be overwhelmed by the information available on their computer screen or on large electronic displays. They might well be attracted by the prospect of trying to analyse the information themselves, rather than leaving this task to specialist staff officers. Higher commanders should be able to employ staff to synthesise and analyse the large amount of information available now and the even larger amount that will be available in the future. It will therefore be a matter of choice or, if you like, temperament and personality, as to whether the commander will allow this additional information to become a burden to him or a valuable tool that will enable him to get inside his enemy's decision cycle. It is certainly a matter that can be improved by suitable training and practical experience. It may also be a source of additional stress.

There is another danger from the explosion of information. The commander might be persuaded to believe that, by sitting in his leather chair in his airconditioned headquarters, he can adequately command his force. He will be able to speak freely with his superiors and subordinates by radio and will probably be able to see them via videoconferencing facilities. His staff and his electronic displays will tell him what the enemy is doing and where his own forces are located. Commercial televisions and displays beamed from UAVs or satellites will give him a general and, in many cases, detailed picture of the various aspects of his campaign. His naval, land and air component commanders will be on hand to advise him. He might also have a political adviser. The commander might think that all these technological advances have been marshalled to reduce his stress, and he might well be right. However, if a commander exercises command in this fashion he is likely to miss the fact that war is a human activity. The commander needs to gain a feel for the battlefield, for the morale of his troops, for the psychological well-being of his subordinates, for the heat and the humidity, for the smell of fear. The idea that new technology might reduce the stress on a higher commander is likely, then, to prove a dangerous illusion.

Physical limitations—the future

It is certainly true that technological developments have made it easier for a higher commander to deal with the physical problems of command. Improved communications enable a commander to exercise control of disparate forces over a large area, while helicopters and executive jets enable him to visit remote corners of his theatre of operations. But physical factors still come into play. Aircraft and surveillance systems are still affected by bad weather. Improved transportation systems have raised expectations about what can be deployed and sustained. The bar is raised ever higher. Commanders will always be pushing the limits of what is possible. Problems of supply and maintenance over long distances and in difficult terrain will continue to be a source of stress.

Political pressure—the future

It is difficult to imagine that higher commanders will come under greater political pressure than they have in the past 50 years, but they probably will. There are a number of reasons for this. Future military deployments are more likely to be for political reasons than for reasons concerned with the direct defence of the country. The deployments will be open to criticism from a public that will receive its information directly by television or on home computers from a variety of international sources. As I mentioned earlier, politicians will have the capacity to deal directly with tactical commanders, bypassing the higher commanders who will still be held accountable. After reviewing the reporting arrangements during the Somalia operation, Breen concluded that it 'would be better to authorise tactical headquarters in the field to report directly to the operational and possibly even the strategic level of command when serious incidents have occurred or are unfolding'.[39] Political pressure will therefore continue to be a prime source of stress for higher commanders.

Control of the media—the future

The control or management of the media will continue to be a major problem confronting all commanders. This problem will continue to grow as new technology enhances the already considerable capacity of the media to beam stories live to television audiences around the world.

Public support—the future

The problem of maintaining public support has become even more critical to the planning and conduct of military operations. This is not just a matter of controlling or working with the media. The commander will need to weigh every operation to determine whether it is likely to affect public support, and he will have to assume that the results of the military operation will become public knowledge relatively quickly. The US air attack on the bunker in Baghdad that resulted in the deaths of a considerable number of civilians, including women and children, presented the Iraqis with a propaganda weapon that they used to strike at public support for the Allied campaign. The commander will need to consider carefully any advice from subordinates that new, technologically advanced weapons will allow a so-called surgical strike with no collateral damage. He will also have to view with scepticism any advice from his intelligence staff that there are no civilians in the vicinity of proposed targets.

Linked to the problem of maintaining public support is the need to take into account the laws of armed conflict. Legal considerations now place commanders under extra stress, especially when the activities of their troops will be open to public gaze. The personal conduct of officers and troops is now under greater scrutiny. Instances of misconduct, ranging from sexual harassment to drunkenness, cultural insensitivity, looting or just plain insubordination are likely to be aired publicly, causing additional stress to a commander.

Cooperation with allies—the future

The problems of cooperating with allies will persist. The collapse of the Soviet Union and the deployment of large multinational forces under the United Nations umbrella has meant that commanders are now likely to find themselves operating with some unexpected allies rather than with the traditional ones. It is unlikely that Lieutenant-General John Sanderson could have imagined in 1987— the year that Presidents Reagan and Gorbachev were still arguing about the INF treaty and Soviet troops were still in Afghanistan—that five years later he would be commanding a force of 12 battalions in Cambodia, including battalions from Bulgaria, Indonesia, Tunisia and Uruguay. It was no easy task.

Host-country authorities—the future

As with the problem of cooperating with allies, the end of the Cold War has given rise to more deployments overseas to countries of non-traditional allies. Peacekeeping operations in locations such as the former Yugoslavia show the complexity of dealing with host-country authorities. Commanders will be faced with difficult political problems involving negotiations with foreign politicians while they try to satisfy the directions of their own government.

Cooperation with allies and dealing with host-country authorities are both cross-cultural activities that introduce an element of unpredictability into command. By challenging expectations and assumptions this element raises the level of stress.[40]

Military jealousy and pride—the future

It would be good to be able to report that this factor has declined in importance, but the experience of history gives us little cause to expect it. It might be hoped that the establishment of the Australian Defence Force Academy, joint headquarters and the increased emphasis on joint training and operations have broken down, and will continue to break down, inter-service jealousies and prejudices. However, strong-minded individuals with a high degree of self-belief will always be found at the top in the military. Without trying to be moralistic, we have become a more self-centred society. Society is more materialistic, and Christian convictions are not so widespread as in the past. We see corruption in politics, business and sport. While we expect higher commanders to display high moral character, we should not be surprised when they reflect existing societal values.

The importance of the campaign—the future

Although is likely that future military operations will be for political purposes rather than for the direct defence of the country, this does not mean that commanding such operations will be less stressful. When Australia was threatened by invasion the problem of defending Australia was shared with many commanders, and indeed the commanders could feel that they had considerable support from the public and understanding from their colleagues. Those commanding deployments overseas with a limited mission in the future might feel that they are largely left to their own devices—that no-one is really interested in their problems.

Managing change

The rate of change in society, including the use of information technology, has accelerated in recent years. Managing change is therefore a fairly new factor. Higher commanders are responsible for implementing change. They might find that old models of leadership are outmoded, that their soldiers have different expectations from those in the past, and that command needs to be exercised through networks rather than in the old, hierarchical fashion. This challenge might prove to be one of the most stressful faced by future commanders.

Ethical dilemmas

Many of the factors I mentioned earlier will conspire to present higher commanders with ethical dilemmas. As a US Army study reported, 'when the costs are high, every risk, responsibility and dilemma is magnified. With the unprecedented political, media, and public oversight in real time in people's living rooms, these significant responsibilities will be more evident and decisions will be even tougher to make'. Commanders will find these decisions extremely difficult and stressful without a strong ethical base.[41]

CONCLUSION

The picture emerging indicates that in the future higher commanders will be faced with increasingly complex political problems. They will have to take note of public opinion and be sensitive to a wide range of outside influences such as cultural differences, civilian authorities and allies. New technology will not necessarily make their job easier, even if it has the potential to do so. There might not be a clearcut enemy. Indeed, in peacekeeping-type operations there might not be any enemy at all.

This does not necessarily mean, however, that future higher commanders are going to succumb to the stress of their appointments. Every age produces its own stresses. It has been claimed that we now live in a more stressful age. We are expected to work harder and longer, be more multiskilled, and accept more demands placed on us. In short, modern-day commanders are products of their environment. It is to be hoped that they will be trained to work with the new information technology.

Command is a very human activity. The traits that enable a commander to stand up in time of stress are those of integrity, moral courage and character. Future higher commanders will face an ever-widening array of challenges, with their attendant stresses.

It is well to remember, however, the words of Sir Basil Liddell Hart: 'Human nature . . . changes but slowly, if at all, and human nature under stress of danger, not at all'.[42]

10 | 'Women as killers and killing women': the implications of 'gender-neutral' armed forces

Eleanor Hancock

IN 1970 NO Western armed forces allowed women in combat or combat-related roles; at the end of the 20th century, some Western countries have no restrictions on women in combat, some have only minor restrictions, while others hesitate before such a historic change in policy. At least one Western navy has a woman submarine commander and one Western army a woman infantry battalion commander.[1] They have taken the first steps towards the equal involvement of men and women in all aspects of the defence of the nation. This process of integration will be a long and continuing one.

Since this chapter was written, some military forces that have been conservative in their policy towards women in combat have moved to adopt, or are being forced to adopt, different approaches. On 4 January 2000 the Israeli Knesset passed a law to open all combat positions in the Israeli armed forces to women. About a week later the European Court of Justice ruled that Germany's ban on women soldiers in combat positions was discriminatory. As a result, the German constitutional ban is to be lifted and women are to be allowed into mainstream German army units from 2001. Italy also has switched from a conscript to a fully volunteer army in which women will be allowed into frontline combat.[2]

The purpose of this chapter is to outline the influences leading to greater use by Western armed forces of women in combat and combat-related roles, the real and perceived problems in this, the willingness of women to kill and of men to kill women, and the possible effect on the conduct of warfare of both sexes formally taking part in combat.[3]

It is appropriate here to make an aside about terminology. The term 'gender integration' is preferable to the term 'gender-neutral' (although this last is the technical term employed) because it is sometimes easy to confuse or elide 'gender-neutral' with 'gender-blind'. Where there has been a history of discrimination, gender blindness can continue the disadvantages to women from previous discrimination.[4]

Given the multidisciplinary nature of this work, reference to historical precedents is limited—the emphasis being on contemporary issues and developments. Accordingly, the discussion of historical examples is restricted to women's combat experience in the Soviet armed forces during World War II. (Another relevant recent comparison is to women's combat in the Vietnam War.)

HISTORICAL EXAMPLES

Today's armed forces are not the first in this century in which women have had access to combat and combat-related roles. In addition to the extensive use of women in revolutionary, guerrilla and partisan warfare, women fought in the Serbian and Russian armies in World War I, for both sides in the Russian Civil War, and for the army of North Vietnam in the Vietnam War.[5]

The most significant recent precedent is the employment of women by the Soviet armed forces in World War II, a precedent that needs far more study. In a supreme popular effort to defeat Nazi Germany, the hard physical labour of Soviet women on farms and working 66-hour weeks in factories under martial law kept the Soviet people fed and Soviet forces equipped.[6] Some 26 000 women had operational roles in the partisan movement.[7] During 1941 an unknown number of others served and died with men in the poorly armed and hastily trained people's militias, which were thrown desperately into battles to help slow the German advance.[8] In 1941 women were already fighting in the front line, despite an existing legal restriction on women to ancillary, medical and specialist roles; by 1942 they were formally admitted to the field forces.[9] Altogether some 800 000 Soviet women served in front-line duties as volunteers in all formations of the Soviet Army, Navy and Air Force; they participated in hand-to-hand combat, and saw action as snipers, infantry, machine-gunners, tank commanders, bomber and fighter pilots, mortar crews, sappers, signallers, artillery gunners and parachutists.[10] (Women in front-line medical roles also appear to have combined these with combat duties.[11]) Women usually served in mixed units, although there were a volunteer women's rifle brigade, three women's infantry battalions and one women's reserve rifle regiment;[12] while some women flew in predominantly male air regiments, there were also three women's air regiments.[13] Women excelled as snipers, and a women's school for sniper training was established.[14] This participation came despite opposition and prejudice on the part of men, officers and commanders, and though little provision could be made for women's specific sanitary and medical requirements.[15]

Women's combat service as guerrillas in the Vietnam War and in the Israeli war of independence may be classified as examples of the greater willingness to deploy women in irregular combat than in regular combat or before a state has been formed.[16] Israeli policy has long since been overtaken by other armed forces. Why, then, do the Israeli armed forces retain a lingering and quite misleading image of gender integration? This remains an image so prevalent that it was the first example many people raised with me. One possible explanation is that for a long time Israel was the only Western country to have partial conscription for women. Israeli Jewish women who are not married, pregnant or strictly religiously observant are called up to serve in the armed forces.[17]

OVERVIEW OF CURRENT POLICIES

Australian policy towards women's roles in the armed forces is under review.[18] Current Australian Defence Force policy is that men and women can compete equally for employment in all positions except those involving combat duties, which are defined as: 'requiring a person to commit, or participate directly in the commission of an act of violence against an armed adversary: and exposing a person to a high probability of direct physical contact with an armed adversary'.[19] This precludes women from employment in Navy clearance diving teams, as airfield defence guards and ground defence officers in the RAAF and—most broadly—in the Army from infantry, armour, artillery and combat engineers' positions. Legal restrictions currently prevent the ADF from placing women in roles involving hand-to-hand combat.[20]

The policies of other Western countries vary widely. They cover a spectrum from countries such as Italy and Germany, which either have no women in their armed forces (Italy) or which allow them only in extremely restricted roles (Germany),[21] through to a middle group (Australia, France,[22] the USA, United Kingdom), which has expanded women's roles in the armed forces in the past 20 years while maintaining some or all combat restrictions. In this group Britain is interesting, in that it has moved in the past 10 years from a more restrictive policy to a policy that is in some ways more liberal than that of Australia, as it allows women in some positions in the artillery.[23] At the most progressive end of the spectrum cluster countries that have in place only minor exclusions or none at all, ranging from Canada (whose only significant combat exclusion—submarine service—may end soon;[24] the exclusion of women from serving as Roman Catholic chaplains is beyond the powers of the Canadian state to change!), to armed forces with no restrictions at all: Sweden, Belgium, Netherlands, Norway, Denmark, Spain and Finland.[25]

The US military has attracted the most publicity and academic attention on this issue because of the size of its military and the greater proportion of women who serve in it; the US role as the one remaining superpower; the international dominance of its mass media; and the fact that the expansion of opportunities for women in the US military has been more fiercely contested than in some other armed forces.[26] This should not blind us to the fact that other armed forces have changed far more readily, dramatically and seemingly easily. The first Canadian woman fighter pilot commented, 'The Americans are still more traditional . . . They're very homemade apple pie—with women at home making it. They think women can only do certain things, whereas in Canada, there was no public protest'.[27] Of those military forces whose policies are more advanced, the Canadian armed forces have women forming a greater percentage of overall members, and Canada appears to be approaching the change in the most systematic manner;[28] the Scandinavian armed forces appear to have been the most successful to date in implementing integration, though on a smaller scale.[29]

FACTORS INFLUENCING THE EXTENSION OF WOMEN'S COMBAT ROLES

What factors have influenced the extension of women's combat roles in the past 20 to 30 years? Some are obvious, others less so. The first factor that has clearly influenced this development is demography. Since the 1960s the cohort of young men available for military service has been smaller than ever before. While demographic patterns have changed temporarily, the most obvious example being the postwar baby boom, the overall demographic pattern in Western countries since the mid-19th century has been a decline in population growth to replacement level or below.[29] Additionally, with the exceptions of the USA and Canada, and until recently Australia, Western countries have been reluctant to turn to immigration to make up any population shortfall. The resultant small family sizes create further problems for Western militaries because they create the phenomenon known by the sexist Italian term *mammismo*—the increasing reluctance of the parents of small families to risk the deaths in action of their children, male or female.[30] Western public opinion is less tolerant of either male or female casualties.[31] This in itself may be a factor strengthening the development of a post-military society.[32] Are the casualties of a volunteer force easier to accept than those of a conscript military? (Does the French Foreign Legion offer a further model of a way for the size and range of use of Western militaries to be extended?)

Second, interacting with both the declining birthrate and the development of a more demilitarised society is a decline in the use of conscription in favour of a volunteer army, as part of a growing professionalisation and managerialisation of the defence forces and as a response to post-military values in the wider society. In continental Europe, there has been a historic change in the policy of many countries since the collapse of communism. While in 1991 all continental European countries, except Liechtenstein, Luxembourg and Monaco, had male conscription,[33] Belgium, the Netherlands and France have since announced its end or suspension.[34] This favours the argument that the potential pool of recruits should be cast as widely as possible, to include both men and women. When armed forces are assured of a continuing pool of young men in the draft, their incentive to expand the role of women may be weaker.

A third factor—the final of the self-evident factors—is of course the resurgence of second-wave feminism in Western countries since the 1960s, which has extended the political, social and economic influence of women, and altered their aspirations and values. The majority of my female students, for example, do not expect to encounter discrimination or prejudice in the workforce. In particular, liberal feminist organisations that concentrate on demands for equal rights, such as the National Organization of Women (NOW) in the USA and the Women's Electoral Lobby (WEL) in Australia, strongly support women's inclusion in combat and combat-related roles.[35] Liberal feminist organisations and activists have stressed, and can be expected to continue to stress, the importance of military service for full and equal citizenship.[36] Because until now liberal feminism has been stronger in the USA than Australia, US feminist organisations have played a more significant role in pressing for change there than they have here. (There is a chance that this may change in Australia, that socialist and cultural feminism may be weakening in favour of liberal feminism.)

The USA too has seen more concerted feminist efforts to influence foreign policy and thus also possibly future defence policy concerning peacemaking and peacekeeping. Examples of this include current campaigns for the recognition of rape as a major war crime as a result of its use in the fighting in the former Yugoslavia, and to withhold de-facto recognition of the Taliban Government in Afghanistan because of its policies of 'gender apartheid'.[37] Note the significance of this last term: it gives the model and intellectual analogy—the campaign against apartheid in South Africa—for the campaign that is being mounted.

The varying policies of Western states on the deployment of women in the armed forces can usually be explained as a result of a complex interaction between the three factors I have outlined. A state whose population is not growing, a state that has a volunteer army and/or a strong commitment to equality, in particular gender equality, or a strong feminist movement, experiences a number of pressures in favour of expanding the role of women in the armed forces. (A further, negative, factor that may encourage such conditions is the need to counterbalance or check other internal divisions within the armed forces: racial, ethnic or linguistic divisions.)[38] Different factors appear to be at work in different countries. I am not sure of the main influences on Canadian or Belgian policies, but it appears likely that in the Scandinavian countries governmental commitment to gender equality was more important than the continuing existence of the male draft. The Netherlands too has a strong feminist movement.[39] In Italy, a country with one of the lowest birthrates in the world, a strong and militant feminist movement[40] has not yet been able to lift a total ban on women's employment in the armed forces. This has occurred not because of active opposition from the Italian military but because of 'slowness, inefficiency and stagnation' in the political system.[41] A commitment in principle to change this policy in the past few years has been delayed by government coalition changes.[42] If the Italian Senate passes the bill currently before it, women will be admitted to the Italian armed forces in the year 2000 and Italy will move immediately from a policy of complete exclusion to one similar to Canada's, with no limitations on women in combat except for some restrictions on deployment on ships and submarines.[43]

In Germany, where women have been constitutionally barred from service involving use of arms since 1968,[44] and where there has been historically strong resistance to the idea of women in the military even in auxiliary roles,[45] women have been allowed in the forces in medical positions and in military bands only since 1975.[46] Ironically, now that German forces are peacekeeping in Bosnia, the constitutional ban has had to be bent to allow women serving as medical officers to carry side-arms for self-protection.[47] The comparative weakness of German feminism and its strong hostility to the military and military service (for men, let alone women) means that there has been little political pressure to change the policy, despite Germany's low birthrate.[48] Feminist groups sympathetic to the strong West German peace movement opposed a 1979 suggestion by the Bundeswehr that women as well as men should be conscripted.[49] (Despite the seemingly egalitarian proposal for gender-neutral conscription, the Bundeswehr proposed to use the women thus conscripted for non-combat duties only.) It remains to be seen whether the greater economic and military role of women in the former East Germany, where young men and women both received pre-military training, will have any effect on the debate in Germany.[50]

A number of other less obvious factors may also contribute to increased willingness to extend the role of women in combat. The fourth factor is the blurring in all forms of 20th century warfare of the distinction between combatant and noncombatant, military and civilian. A greater percentage of casualties in 20th century warfare have been civilian than in previous centuries.[51] This has weakened considerably the argument that civilians are somehow protected from war.

A fifth factor may be the greater evidence of female aggression and use of violence in society, which makes it less unthinkable to see women in combat roles. (In this context, I am not suggesting that military effectiveness is the result of personal aggressiveness.) Analysts disagree on whether women have become more violent as they have become more emancipated, or whether female emancipation has led to a greater readiness on the part of society to recognise violence by women that has always existed.[52] Whatever the reason, rates of violent crime in women have been rising since the 1960s.[53] In addition, women's role in political violence has been highlighted by the participation of some women politicians in inciting—and of ordinary Hutu women as well as men, including allegedly pregnant women and nuns, in committing—genocide in Rwanda in 1995.[54]

All of these developments may create a sixth factor: an expectation on the part of some young men that women should do their bit, particularly if a draft is ever reintroduced in Anglo-Saxon countries.

The seventh influence that may develop is the political and funding cost to the military of opposing these policies.[55] There is not only the advantage of maximising the range of choice for all positions,[56] but the political dangers of a military too unrepresentative of society in its composition and values. Armed forces in a democracy should be representative of the society they defend. (Employment opportunities for spouses of defence personnel and ensuring a representative mix of ethnic backgrounds are other important areas for the Australian Defence Force.) Greater involvement of women in defence forces should also contribute to overcoming the 'gender gap' in attitudes to defence and foreign policy issues and funding.[57] Equally, why should women as taxpayers support an institution that discriminates against them?[58] How effectively will officers and commanders from armed forces that restrict access to combat roles by gender work with their allies from gender-neutral armed forces?

OBJECTIONS TO WOMEN IN COMBAT

In the course of this research I have been asked to list the problems and objections that have arisen to the expansion of women's roles in combat and combat-related positions. I am sure that others can think of potential problems that have not yet occurred to me. The Soviet case suggests that, when integration *has* to work, ways are found around many of the anticipated problems. Men and women bonded in small fighting units. By cooperation, men and women accorded each other privacy when necessary. In a society where women were expected to, and did, undertake hard physical labour in the paid workforce,[59] women were capable of the acts of physical strength required in combat.[60] A quotation from a 1943 poem by the soldier-poet Georgy Suvorov illustrates the question of strength:

> Oh! How strong you are,
> A woman's hand on a machine gun![61]

Once again I want to note the major objections and some, less frequently mentioned, difficulties. While the major objection has stressed women's physical weakness, particularly their comparative lack of upper-body strength, the other significant difficulties raised include the effect of women on building and maintaining group cohesiveness, and the overall social and psychological taboos and prejudices that need to be overcome. A recent Canadian study has suggested that this last factor—male opposition based on self-interest, taboos and prejudice—is the most difficult, and that insufficient upper-body strength is often an ostensible objection that in fact conceals these wider prejudices.[62] The physical standards required for Canadian combat arms were often found to be ambiguous or difficult to justify.[63] Nonetheless, upper-body strength is a real problem for gender integration in the armed forces. Men are on average 10 per cent taller, 20 per cent heavier and 30 per cent stronger, especially in their upper bodies.[64] Women are also on average more likely to have knee injuries in heavy physical training, unless they take preventive measures—strengthening the knee by targeted extra exercises.[65]

For women to complete military training requiring upper-body strength, two new measures will probably be necessary. First, they may need selection based more specifically on their physical capacity than currently appears to be the case in some of the armed forces; second, they may need a specific targeted pretraining program to build up their strength and endurance.[66] US Army studies suggest that with such a program they are capable of completing combat training without standards being lowered.[67] (I am not suggesting that the necessary standards should be lowered.) Canadian experience suggests that a 'critical mass' of women on combat training courses is necessary. Both men and women resent the focus on a small number of women when they are pioneers, and as a result women feel that they are being micromanaged, often to their detriment—this 'critical mass' must have the physical capacity to complete the training.[68]

Physical training in most Western militaries appears to emphasise those areas where men tend to excel and not to test or train those where women tend to excel. Can military forces in combat training make better use of women's average greater stamina and increased capacity to endure extreme temperatures and extreme conditions in general?[69] Soviet accounts noted that women often finished long marches of 30–40 kilometres in better condition than the men.[70] Nazi commandants in the extermination, concentration and labour camps saw women prisoners as more resistant to extreme conditions than men.[71]

A second difficulty is the bonding issue. Does the military need to recognise that bonding between women, as well as between men, is important?[72] Has a military persona that will comfortably fit women been devised yet?[73] Is an ethos of professionalism enough to bond disparate groups from the same society?[74] Building and maintaining group cohesion in a mixed-gender fighting group may need to rely on different mechanisms than in an all-male group; this change in itself—finding out what will work and what will work best—may take some time.[75] Soviet experience suggests that it can be done. It may be easier to achieve in combat than out of it. US experience of racial tensions in Vietnam suggests that, even when there were severe racial tensions away from the front, racially mixed groups bonded when they had to—under fire.[76]

Perhaps the third and greatest difficulty is the social taboos, prejudices and self-interested objections that influence opinion both inside and outside the military, and which have led some members of the armed forces in both Canada and the USA either consciously or unconsciously to sabotage or undermine policies for greater female access.[77] It is their persistence that has led a recent Canadian study to see the upper-body strength issue as 'the curtain that masks other issues'.[78] Where women genuinely met the physical standards, the Canadian study found that some of the men involved refused to believe this, discounted the standards and women's success, and immediately switched to questioning the sexual activities and/or leadership of the women involved.[79] Judith Hicks Stiehm in *Bring Me Men and Women* has written perceptively about the wide range of taboos, emotions and prejudices that come into play.[80]

At the end of the 20th century, which has seen *inter alia* deliberate bombing of civilian targets and a fluidity in warfare that requires support troops to be close to the front, it is astonishing to find objections to women in combat that still derive from a belief that excluding women from combat protects them from the dangers of war.[81] Ill-treatment of prisoners of war in general, and of women either as prisoners or enemy civilians, is *not* dependent on whether they have been combatants or noncombatants. Japanese soldiers in World War II did not discriminate between nurses and troops when they shot the survivors of the *Vyner Brooke* on Banka Island in 1942.[82] Until April 1945 Nazi Germany was punctilious about keeping women away from any formal or legal military role, preferring to call up reluctant 60-year-old men than 20-year-old women,[83] but their civilian status did not protect the thousands of women tortured and/or killed by advancing Soviet forces, the tens of thousands of German women raped by Allied troops, and the tens of thousands deported to forced labour in the Soviet Union.[84] Indeed it can be argued that this experience created an hostility to, and devaluing of, the military in Germany for generations. In 1945 German men in and outside the military failed to deliver on the promise to protect women which had implicitly and explicitly justified their male privileges—and the military in Germany lost a prestige it has not yet regained.[85] Civilian or noncombatant status leaves women even more defenceless.[86] At times fighting forces have given better treatment to their fellow enemy soldiers than to enemy civilians.[87] Might women in the forces at times be more vulnerable if they are not trained to defend themselves as well as possible,[88] and if they are dependent on men's decisions on whether they will continue to be protected or not? This is an argument for more training of all women in the armed forces in combat techniques, not less.[89] At least, to use the slogan current in Britain in 1940, you can always take one with you!

Equally, though men do not generally speak of it and may not be as aware of it, rape is not necessarily directed only against women. The extent of rape of men in Western society is probably underestimated and underreported. Men captured by the Iraqis in the Gulf War feared rape and sexual assault.[90]

At times the idea surfaces that women need to be protected as breeding stock necessary for the nation's future.[91] It would seem that this is an argument rather for young men and women equally to run the risks of dying for their country, to protect the breeding balance between the sexes. Despite artificial insemination and single parenthood, past experience (in Germany, Austria and the USSR after World War II, for example) suggests that large-scale losses of young men, even if young women are comparatively protected, will still lower the birthrate. If men do not survive as well, who will be the partners of the surviving young women? Arguments about protection of women from war, or war's effects, by excluding them from combat, are based on an outdated view of war.

There is a range of other potential difficulties, some of which come from the fact that we are now in this period of transition. These include problems related to sexual modesty (for example, mixed-gender bathrooms and living quarters), which have been a great problem in the US military. A colleague of mine argues that this is a combination of the sexualisation of US culture and its strong homophobia. The Scandinavian armed forces, coming from cultures more matter-of-fact about nudity, have not encountered these problems: men and women share bathing facilities and mixed accommodation in the Danish and Swedish navies, for example.[92] Difficulties include hostility and opposition from women who joined the military not expecting to be required to go into combat.[93] There is also the problem of hostility arising from those individuals that oppose change emanating from 'social engineering'. Is the military right to fear men's protective instincts coming into play in combat, or are these impulses socially constructed?[94] Do volunteer militaries disproportionately attract men who are strongly attached to traditional sex roles and women who are not,[95] and does this difference cause problems for gender integration? Integration in some areas of the military may be easier than others: integration has progressed further in the Swedish Air Force than in the other services, for example.[96] There is no doubt that issues like fraternisation (which none of the current approaches seem to handle satisfactorily), sexual harassment,[97] policies combining readiness with pregnancy and family commitments,[98] the integration of training[99] and the need to change training styles all are as yet unresolved.

These issues appear to have been greater problems for the military in Anglophone countries than in Scandinavian countries. Several questions arise: Are these problems a reflection of the relative size of the militaries involved? Have the Scandinavian militaries been more successful because they have taken integration for granted? Are there other cultural factors involved? Is this a result of the greater public and economic role played by women in Scandinavia for decades? Is it possible that integration for combat has worked, particularly in the Scandinavian military, precisely because they have retained male conscription? What do I mean by this suggestion? Career officers and NCOs in these armed forces may be used to training and motivating a far broader cross-section of modern male youth than is the case in the volunteer armies of Anglophone societies. They may as a result have to be more flexible and adapt to a wider range of social attitudes. Because these armies are based on conscription, the proportion of men seeking to validate their masculinity through military service may be smaller.[100] And if officers and NCOs are used to training a wide cross-section of drafted young men with varying degrees of commitment and physical fitness, instead of a self-selecting group, might female volunteers that have chosen to be there be better or easier to train?[101]

In those integration efforts that have been the most studied, in the US armed forces in particular, there can be no doubt that many mistakes have been made that have magnified the difficulties—from inappropriate and often unnecessary changes in the standards required to mistaking the difference between real and perceived fairness. Some of these arose out of opposition to the new policies.[102] In the US many of the steps that have been taken to ensure and maintain the success of the racial integration of the armed forces, and to monitor equal treatment for male African–American soldiers, have not been taken to ensure the greater success of women's opportunities in the forces. Above all, the widespread belief that it cannot be done—it will not work—can in itself be a self-fulfilling prophecy. Do you get maximum performance out of anyone by telling them from the beginning that they will be too weak and incapable to do many aspects of the job?[103]

My question when we consider the difficulties is: if we want integration of women in combat and combat-related duties to work, if it has to work, how do we make it work? There were, for example, a number of factors that may have assisted in the successful Soviet employment of women in combat in World War II, including the existence in Russia of a female warrior tradition; an ideology and society that preached gender equality, even if this was not always practised; the premilitary training given to women as well as men in the 1930s; and the fact that many women volunteers came from a highly ideologically committed group—the Komsomol.[104] In contrast, in Nazi Germany measures to make greater use of women to assist the armed forces were taken with consierable hesitation and opposition. German women had been given no premilitary training; Nazi ideology strongly supported gender inequality; and it is unsurprising that last-minute attempts to use women in combat had little success. The regime could not successfully tap the strong commitment of some young women to Nazism.[105]

THE FUTURE FOR WOMEN IN COMBAT

What needs to be done? Will we need different policies to make it work for a larger cohort of women than the small group of pioneers that will be the first to succeed?[106] Do we need to recruit women differently, picking them more specifically for their physical capacity? Are military forces that are less tolerant of homophobia more likely to succeed? Homophobia unfairly affects *all* women in the forces that do not fit conventional feminine stereotypes—whether they are heterosexual, bisexual or homosexual—and therefore it affects many women most likely to succeed.[107]

Have serious policies been adopted to make integration succeed? As General Colin Powell has remarked, the success of racial integration in the US military over the past 50 years required the implementation of specific policies. These measures illustrate that, for integration to succeed, specific policies have to be implemented. In 1948 a degree of social engineering was involved in integrating the US armed forces. This level of social engineering has so far not been attempted for women.

In addition to President Truman's setting up a commission to ensure that there was no backsliding by any of the forces on desegregation, officers were held accountable for any problems, and non-compliance was treated as a command failure. Once the armed forces were formally integrated, integration of actions and practices followed. This included a policy of no tolerance with respect to expression of racist beliefs in any manner.[108] (The absence of enforcement of no tolerance of sexist statements is a major complaint in both the US and Canadian militaries.[109])Racial integration included education to explain difference; it included, and still includes, monitoring of the distribution of promotions, punishments, assignments and bonuses by race; outlets for African–American soldiers and officers to complain of racist behaviour outside the normal chains of command; raids on officers to monitor the racial climate of their units; and an equal-opportunity rating for all officers in their efficiency reports. Unacceptable ratings meant the end to an officer's career. Commanders were provided with training and specialists to help them. African–American officers were recruited initially mainly from graduates of the historically black colleges and universities. It was not until the 1970s that superintendents of military academies were set goals of percentages of minority students to recruit. If they needed help to meet the academic standards of the military academies, minority recruits were sent to a one-year preparatory school to up-grade their skills. (Might some female recruits benefit from a similar preparatory program for physical fitness?) Minority officers were mentored and their promotion patterns monitored to remove patterns of discrimination. Despite all these steps, minority officers still do not get promoted at equivalent rates to their white counterparts, and there are continuing racial problems in the US armed services. This is evidence that such programs need to be continued long after the problem appears to be solved.[110] In comparison, only some of these policies have been introduced for women in the US military, and it is perhaps not surprising that there has been far less success. Sexist attitudes, for example, are still openly expressed towards women in the forces where racist attitudes have been driven underground.

During my research I was asked to consider the questions: will women kill, and will men kill women? It had never occurred to me that we even needed to ask this latter question. History and the treatment of women in many parts of the world suggest that this of all issues is one about which military policy-makers should not worry. Many men do not appear to have a problem killing women—at least killing unarmed women.[111] In fact, some feminists have coined the word 'femicide' to characterise male killings of women as a means of keeping women in their 'rightful' (i.e. subordinate) place.[112] Christopher Browning's research has shown how easily 'ordinary men' could be brought to kill unarmed Jewish women and children. When it is a matter of life and death in combat, I do not believe that men will hesitate to kill their female enemies. Nor should they. Both men and women may need different training to overcome any conditioning against attacking and killing the opposite sex, but I am sure that this can be developed.

Will women kill? For obvious reasons it is difficult to set up an experiment to prove this: a repeat of the Milgram experiment, testing women's willingness to administer fatal electric shocks, suggests that women will do so.[113] The nurses that helped in Nazi Germany's forced euthanasia program and the female concentration camp guards drafted into the SS as part of their labour service were ordinary women, not specially selected for their brutal roles. We possibly underestimate the extent to which women do kill in Western societies. What evidence there is from partisan and guerrilla movements, terrorist organisations, the SOE and the Soviet armed forces suggests that women will kill.[114] Soviet women snipers found that killing Germans became an everyday occurrence.[115]

What effect will this have on the conduct of war and the fighting environment? The one thing the historian learns is never to predict. The strategic situation in our world is shifting as we speak. Who could have foreseen two years ago that we would be contemplating the prospects of independence for East Timor, for example? What will be the shape of the military-after-next? In the post–Cold War world the situation is, as the Swedish armed forces' homepage puts it, one of 'fewer threats but less peace'.[116] There are challenges from 'rogue states', low-level conflicts, and in peace-making or peacekeeping.[117] At the same time armed forces must prepare for a possible major war involving high technology.

This will require a considerable degree of flexibility in military leadership and training. Does the current education and training in armed forces foster this flexibility, either for men or women? Male reservists in the Canadian armed forces, for example, have described the regular Canadian forces as neither flexible nor pro-gressive but rather rigid and 50 years behind the times.[118] Women might be able to contribute to this flexibility,[119] but they will be less likely to do so if they are a beleaguered minority in the military. Will women change the military, or will the military change women? The balance of historical experience suggests that the latter will be the case, at least initially.

Men and women in Western armies may find themselves fighting alongside allies with different values on gender issues, as in the Gulf War; they may fight, or peace-keep, between nation-alist, religious fundamentalist (particularly Islamic) or populist guerrilla forces as well as more conventional forces. If Western armed forces find themselves involved in low-level conflict, they may well fight forces that use women in combat. Some women fought in the various irregular militias in the former Yugoslavia. Will it be possible to use a commitment to gender equity to undermine some Islamic states, by appealing to a potentially very dissatisfied group in these states—women?[120]

Hugh Smith has argued that the exclusion of women from formal conflict has been the greatest human act of disarmament.[121] If so, perhaps it also illustrates the costs of unilateral disarmament.[122] He suggests that without it, states might be able to escalate warfare to a previously unknown level of intensity. This may be so—certainly the USSR could not have survived 1941 and 1942 without its ruthless determination to use the resources of its entire population, and without the response of Soviet women.[123] Use of women in combat in time of crisis can be seen as a sign of desperation.[124] (In passing I might note that similar fears were expressed concerning the racial integration of the US defence forces—that this might be seen by allies and potential enemies alike as a sign of weakness.[125]) Alternatively it can be seen as a sign of determination to win, if necessary by making every sacrifice. If it is seen as the first, it may encourage the enemy; if the second, it may dishearten them. A variety of possible enemy reactions have been posited. Will a male enemy force fight even harder and be less likely to surrender because this would symbolise, at least to some of them, the ultimate loss of masculinity?[126] It is not yet clear which, if any, of these responses is likely. (Of course, one reaction does not preclude another.) Did US and Australian forces fight better in Vietnam because they faced female as well as male adversaries? As far as I am aware, no-one has argued this point. Equally, the brutality of the fighting on the Eastern Front during World War II, and the frequent lack of quarter given to surrendering troops, occurred for reasons other than, and before, Soviet use of women in combat. German forces and later Iraqis have surrendered to armies containing women.

CONCLUSION

We cannot yet therefore be certain how the use of women as well as men will transform the battlefield. My feeling is that, as has generally been the case when other sectors of Western societies have been integrated, the change will not be as great as we expect, hope or fear. I began by noting that we are at the beginning of a period of transition to achieving integrated armed forces in Western societies. In most Western societies it is less than a hundred years since women achieved equal political rights; some aspects, such as full citizenship rights, have come still more recently.[127] Military service is seen as an integral and normative part of citizenship.[128] In the 19th century as political rights were extended to all male citizens, women's exclusion from service in the armed forces was used to justify their exclusion from political rights. General Powell has spoken eloquently of black men's willingness to shed their blood for the USA, which helped win and secure their citizenship rights. Women are still on what has been called 'the uneven and unfinished path to women's full citizenship'.[129] The process currently under way of opening combat and combat-related roles to them will be part of its completion. Success will require a commitment equal to that demanded of, and given by, the US Government in the racial integration of their armed forces.

11 | Human factors in field training for battle: realistically reproducing chaos

Jeremy Manton, Carlene Wilson and Helen Braithwaite

COMBAT TRAINING HAS over the past 20 years undergone significant change. Technology has permitted a greater degree of data collection on combat team activity than ever before, with the consequence that event data can be recorded, analysed, discussed and fed back to participants faster and in more objective detail than ever before. Over the same period many scientific and management tools have been developed that permit the generation and dissemination of knowledge, doctrine and purpose.

The results of combat team training have demonstrated significant objective improvements in military effectiveness in exercises. However, the cost of achieving this improved performance is high, and there is apparently no objective evidence that performance in land exercises transfers to performance in war. War is characterised by the features of danger, friction, chaos and uncertainty. Clearly, skills acquired in combat training need to be robust to be expressed in war. This chapter examines some of the available evidence on this matter, particularly human factors in team training for combat.

A major theme that runs through military studies is the impact that experiential learning has on combat team performance. Notwithstanding the impact of technologies on military capability, it is a soldier's ability to learn and the speed with which he does so, both in training and on the battlefield, that will affect battle outcomes. Acquisition of military skills and the ability to implement them in the chaos of the battlefield is a key to success. The psychology literature concerned with stress and its effect on skill acquisition and performance is reviewed here.

New technologies available for tactical engagement simulation (TES) afford us the opportunity to conduct more realistic combat team training and develop better-prepared and ready fighting-forces. The growing capability to measure combat performance and the need to test the effectiveness of new weapons, information and organisational systems underlines the sympathy between combat training and advanced warfighting experiments. The issues of simulation and combat training are examined here in respect of future directions in combat training.

THE TRAINING AND EQUIPMENT TRADE-OFF

Conclusions reached by Deitchman, after a close evaluation of the available literature, indicate that:

- Realistic field training in TES improves unit performance by a factor, on average, of two. Military performance improvement due to concerted training is roughly equivalent to the intro-duction of a new generation of equipment.
- The cost of training and maintaining performance at a peak level is twice that of the life-cycle cost of the equipment. The life-cycle cost of a major equipment—for example, a major weapon system—is 1.5 to 2.5 times the capital cost.
- Once performance has been improved, personnel turbulence and skill attrition will cause degradation in performance.
- Automated components of new systems can improve perform-ance and allow training to concentrate on higher-level system performance issues.[1]

As a consequence of such findings, the US Army directed more funds to combat training at the expense of other items in their operations and support budget.

AN EVOLUTION OF COMBAT TRAINING

In the mid-1970s the US Army undertook to assess the effective-ness of the 5-tank and a new 3-tank platoon concept. The 3 by 5 Tank Platoon Test was conducted using a laser-aided engagement system in a live-fire defensive action where enemy tanks out-numbered defenders 4:1.[2]

To control for leadership, experience and lack of doctrine for the 3-tank platoon concept, an interesting experimental design was implemented. At the end of week one, the 5-tank platoon sent two tanks back to the parent unit and became the 3-tank platoon for week two; while the 3-tank platoon of week one was reinforced with two new tank crews and became the 5-tank platoon for week two.

Results demonstrated that in week one the 5-tank platoon performed slightly better than the 3-tank platoon. However, in week two the 3-tank platoon was significantly better than the 5-tank platoon. The 3-tank platoon progressed from losing all its tanks in week one to killing all attackers in week two without a single loss. This finding did not help determine which tank platoon size was more effective, as the results had been confounded by the impact of experiential learning.

The value of experiential learning for combat performance was tested in a controlled comparison between traditional training techniques and the TES approach. The US Army Research Institute conducted a transfer of training study, where small units were evaluated under controlled conditions.[3] Units were assessed on a pretest to determine baseline performance (using TES methods). Half the units were then given traditional training that stressed doctrine, principles, terminology and procedures, and conventional methods were used to verify observance of what had been taught. The other units were given TES training that stressed learning by doing. This regimen was less didactic, with free manoeuvre and casualty assessments providing feedback. Most of the learning with TES occurred in a frank after-action review (AAR) procedure. In the post-test phase each unit was given a mission to be accomplished against an expert opposing force and against other participating units.

The results determined that infantry rifle squads improved overall by a factor of two when they experienced TES training. Conventionally trained squads demonstrated no change and in some cases slightly worse performance.

The research was replicated with armour and anti-armour units and similar conclusions were reached. A significant point made about these trials was that the initial pretest performance of the test units was deplorably low.[4] From these initial studies the notion of developing a more formal TES capability was developed.

THE VALUE OF AN EVALUATION PROCESS

Research at the US Army's National Training Center (NTC) determined a need to improve direct-fire control performance of companies.[5] A research program was set up to examine the problem. Data were collected between 1993 and 1994 on company teams during the planning and execution of direct-fire engagements to determine: areas where companies performed well and where they needed to improve; whether companies improved in fire control during a rotation; how well companies that developed good plans were able to execute direct-fire control. Multiple observations were made on 52 commanders. These data facilitated analysis of how their companies improved over battle serials during a rotation. Hallmark and Crowley concluded that:

- Companies are better at planning activities than executing them.
- Overall execution performance is poor.
- Direct-fire control is not performed as well as movement and positioning.
- Basic planning activities are performed adequately by most companies; complex planning activities are not.
- All categories of activities, except direct-fire execution, improve over battles.

Complex planning is defined as 'planning associated with visualising the way battles develop'.[6] Few commanders reacted well to enemy fire and movement. In the planning phase, they appeared unable to integrate terrain, enemy and friendly factors into an idea of how the battle would flow. As a consequence, commanders were less likely to develop a tactical plan that would succeed.

Findings indicated that tactical commanders generally lack an *integrated skill* associated with insight into the flow of battle. This conclusion was juxtaposed with the finding that companies improved their performance, from a generally low entry-level performance on most measures, during the rotation at the NTC. However, they did not improve in direct-fire control execution.[7]

Performance improvement was noted for command functions as well as platoon and subordinate element synchronisation. Improvement was significant for most variables rated and to some extent attributed to observer/controllers (O/Cs) spending considerable time coaching their counterparts on the *cognitive skills* of battlefield visualisation. It was recommended that these skills be trained at home station. Simulation, integrated with field training, could provide the necessary stimulus conditions for skill acquisition. The studies mentioned by Gorman and the Hallmark and Crowley study represent major undertakings, and the data treatments are sophisticated. While this work was conducted within the US Army's enterprise culture, it demonstrates the value of an objective method in establishing propositions that can be audited.

STAFF MANAGEMENTS SKILLS

Gorman noted that commanders needed good staff teams to support them. This was emphasised in cases where a commander that was expected to perform well in a tactical engagement simulation in the field performed poorly, and a commander that was expected to perform badly did well. This was attributed to the fact that the commander in the first case had not formed a team in his command group, and, in the second case, a competent command team was able to carry a plodder. Good organisational ability of staff teams is emphasised by Deitchman as a quality of effective leadership.[8] Thompson and co-workers indicated that synchronisation of critical command and staff activities was a significant issue in performance.[9] Data from Hallmark and Crowley indicate that company command teams did not manage available preparation time adequately.

Skill and skill acquisition

Psychologists have formally studied skill and skill acquisition since the 1890s, and observed the qualities of skill behaviour to be somewhat enigmatic.[10] We know that skill is a capability that is flexible and adaptive to the environment that the operator perceives. Skill takes time to acquire and is directed by a criterion of ideal performance. Once a skill is attained, one can generate virtual feedback that can make it appear that the action has an open-loop relationship with the environment. Skilled performance is parsimonious of effort, is a pleasure to perform, and is robust to stress and strain. Skills are acquired through practice, and performance decreases when practice is not conducted for some period of time. Skills are acquired and maintained by an active, goal-directed interaction with the environment (tuning).

Much of the research that led to these conclusions emanated from the information-processing and cognitive-psychology literature, and many of its tenets have been recast in the construct of 'situation awareness" or 'SA'. SA has emerged principally out of the aviation accident and nuclear power industry research and bounds the problem space of how to represent the issues of perception, decision making and action in complex environments.[11] Endsley poses three levels of SA: level 1 is the perception of status, attributes and dynamics in the environment; level 2 is the understanding of the significance of these elements in the light of pertinent goals; level 3 is the ability to project the actions of the elements in the environment into the future.[12] This level 3 feature was postulated by Welford in his model, which contends that a skilled operator could simulate, cognitively, the outcome of a proposed action and use the output of the simulation to modify future actions.[13]

Geiselman and Samet used schema theory, in a laboratory study, to analyse staff officer situation appreciations at corps level. Their results can be reinterpreted against the SA three-level model, with the assertion that poor appreciations represented level 1 SA and were characterised by a static portrayal of the enemy situation. Good appreciations, characterising level 3 SA, emphasised the dynamic aspects of the enemy situation.[14]

Naturalistic decision making (NDM), a relatively new and related field of research, focuses on the way people use their experience to make decisions in field settings. This domain of research grew out of a reaction to the traditional decision making studies that specialised in artificial and contrived decision making paradigms, which seemed to bear no relation to real-world problems. One of Klein's observations about NDM was that experienced fireground commanders rarely compared the merits of alternative courses of action.[15] Analysis indicated that skilled commanders are able to choose a course of action, based on the recognition of a pattern of elements in the environment—that is, related to a strategy they have experienced already. This process, termed recognition-primed decision making (RPD), is a construct that proposes that skilled commanders perceive patterns of events, entities and states that evoke strategies of response. These responses are based on previous experience from which they can run a mental simulation of the outcome of the action they will perform. Pattern recognition allows a rapid perception of the situation. Experienced commanders tend to construct complete and coherent situation models.[16] It is worth noting that Geiselman and Samet found that good appreciations by corps-level staff officers did not leave gaps in their summaries and included comment on what key information was not known.[17]

The research theories of skill, SA and NDM, provide rich sources of theoretical constructs that can be used to analyse command team performance and meta-cognitive strategies for developing skills for improved team processes and battle flow visualisation.

A HIERARCHY OF REAL-WORLD COMPLEXITY

It was noted at the NTC that company commander functions, as well as platoon and subunits synchronisation, improved over the seven battles.[18] This indicates that emergent functional issues require specific skill acquisition as higher levels of collective systems training are conducted. These emergent skill requirements are not readily open to individual and collective consciousness until the higher-level activity is experienced along with the appropriate context.

Battle skills take practice to develop, and new elements of the skill are learnt with each situated experience. With complex skill acquisition, new awareness may be perceived for the first time after many iterations of the key activity. Summing up the experiences of the soldiers of the First Australian Task Force on departure from the Area of Operations Surfers, north-east of Saigon in June 1968, Lex McAulay indicated that:

> The men of the infantry, artillery and tanks, both as groups and as individuals, came out of the operation with a deeper appreciation of the realities of war. The soldiers now understood the reason for many of the lessons and the points the more experienced members had tried to make . . . After it, there was much more professionalism, a more serious acceptance of the business of soldiering.[19]

The issue for the combat-training scenario-designer is to create events that will generate the appropriate learning experiences for a unit. In battle the objective is to win; in combat training it is to develop the skills and techniques that in battle will overwhelm the enemy. The accent then is on learning how to fight.

The US Joint Simulation System (JSIMS) is a large, techno-logically focused, simulation system that is being developed to provide realistic joint training to the ready forces, across all phases of military operations, for all types of missions. In 1998 the JSIMS Joint Project Office balanced the technical–engineering focus with the formal effort of a Learning Methodologies Working Group to influence the design from a learning perspective.[20] The centrepiece of the effort is the 'learning methodology process model'.

This model is a fusion of learning theories and practical experience (some enshrined in training doctrine) from each of the services' training environments. The reference document presents guidelines for the conduct of combat simulation and the require-ments for specification of a systematic group of variables. The accent is on tracking the occurrence of events and collecting data, integrating ground truth with perceived data, and performance measurements. The focus is on battle flow as a process, which needs to be articulated and represented as a series of events that can be identified, measured and assessed. Feedback on performance is then provided to participants.

Bondanella et al. reviewed US Army staff training for combat service support and concluded that training was oriented towards crisis response and solving individual problems.[21] Simulation-based exercises were unrealistic and too expensive. Training focused on segments of performance in stovepiped functions, without a view of how output affected other segments of the process. The research group recommended that the army move away from a functionally driven, project-oriented approach, where performance is measured on how well a specific subsegment of the process is performed. A process (activity) measure approach moves the organisation towards a broader and more integrated systems view that is more cognisant of the effects that key activities have on the upstream and down-stream dynamics of the enterprise process and performance. In this context, process measures are not static; they continually change to reflect changes in the environment and continuous improvement.

COMBAT TRAINING AND FIGHTING

The relationship of combat-training performance to battlefield performance is a key issue in the rationale for developing combat-training centres. In the UK, Rowland conducted an analysis of performance data collected during urban fighting trials in the Kings Ride series, a set of trials at a research establishment and trials at the Ruhlben training area in Berlin.[22] The exercise data show a significant improvement in performance (reduced numbers of losses to the attacker) over attack serials. Rowland was also able to compare performance over successive battles from historical wartime data. Using a series of statistical normalising techniques, he determined that units', where there was little or no replacement of own casualties, experience of battles improved performance at a similar rate to training serials. This is notwithstanding the finding that performance in real World War II urban battles was 3.8 to 7.2 times worse than in trials.

Rowland also points out a significant difference in casualties between simulated and real combat. In simulated combat all defenders fought until they were 'killed', and in live combat 20 per cent were killed, 60 per cent were wounded or captured, and 20 per cent escaped or withdrew.

Horowitz et al. examined the link between the training of US Army, Navy, and Air Force units and performance during the Gulf War.[23] They found that the US Department of Defense did not collect data concerning unit performance during the war. No data on how the type or amount of training prior to deployment did (or did not) contribute to combat effectiveness were available.

MEASUREMENT OF PERFORMANCE IN COMBAT TRAINING

The focus of combat-training development in the Australian Army is the evaluation of collective combat performance in realistic and safe conditions through immediate feedback and mentoring.[24] The key feature of the training model is the provision of information concerning performance that can be fed back to participants as quickly as possible after the training event. Assessment cards filled out by the O/Cs at the NTC and those used at the Australian Interim Combat Training Centre assess unit activities during the battle by determining whether a task behaviour occurred or not and, on the O/C's assessment, how well it was performed. This process assessment model is diagnostic and as objective as feasible.

The level at which one collects and analyses activity data on the training battlefield will be determined by a judgement about what to train and how to measure performance in relation to the goal. The US Army and its researchers have been through a number of iterations on how to gather and analyse data on combat-training performance. As training needs change with experience and national taskings, the need to analyse and revise one's methodology is apparent.

Lewman reports that the measurement activity started at the NTC by focusing on the battalion taskforce, and data on the company teams and platoons were included as they were integral to the battalion.[25] In the first instance, a taxonomy of mission-critical tasks was identified and refined. Protocols for measurement, standards and mission conditions were developed along with a battle flow framework and mission segments. Task sequences and linkages between mission segments and teams were developed. Over time, it was determined that this fine-grain information did not provide useful indicators of mission performance. Aggregation of task data at battle operating system (BOS) level lacked sufficient detail to prescribe a training program for a unit to perform a mission-essential task. A task analysis was carried out to identify the critical combat functions (CCFs), in order to design a training strategy for a battalion combined-arms taskforce. These CCFs provide a moderate level of aggregation between mission training plans and BOS profiles. Evolution of the NTC measurement and feedback system occurred over a decade. This emphasises the emergent nature of issues to be addressed as one gains experience with the combat-training process.

There are significant opportunities to develop micro-instrumentation of a company, from the commander down to the soldier, in the combat-training centre (CTC) battle and to gain accurate data on a range of individual and team actions. Fusion of these data with ground-truth indicators will give accurate information on task activities and performance. The accuracy and timeliness of the feedback are the key aspects to modifying combat team behaviour. There will always be a need to have an O/C participate in the process to manage the scenario and tune it to the capability of the company in training.

Process tracing and assessment techniques for home base simulator-enhanced training have been developed and have useful diagnostic attributes. McClusky et al. developed TRACES (tactically relevant assessment of combat events) for an army-training program that is very similar to the NTC assessment method.[26] Initially, a training needs survey identified critical tasks, which were then mapped onto a mission events list (MEL). Observable and acceptable behaviours were then identified by subject matter experts *a priori*, and O/Cs were required to observe exercise behaviour and rate whether an activity was present or absent. The results indicated improvements in performance for an initially poorly performing unit and provided good diagnostic information.

INVARIANT AND EMERGENT ASPECTS OF WAR

Warfare is competitive. Leaders and tacticians aim to overwhelm the enemy or use tactics that are unexpected and that exploit the weaknesses in the opponent's expectation or posture. A key element of warfare is the ability to gather information on the enemy's approach and doctrine, and to adapt one's own posture and practice to achieve an advantage. Ability to learn from the exchanges that occur on the battlefield and the speed of acquisition of knowledge and adaptation to it is a key to success.

Doctrine and the wealth of published literature on the analysis of military operations have defined the invariant characteristics of successful warfare. Australian Army doctrine reframes these invariants against the current climate of military–political thinking and emphasises the tenets of the learning organisation, with its accent on adapting to the emergent situations during combat. Combat training must, therefore, reinforce the acquisition and retention of the invariant principles of warfare and develop the skills to exploit the emergent situations of combat.

Invariants of good battle processes are defined through corporate memory, individual experience, research and feedback. Operations analysis, based on the rational treatment of subjective and objective data, does much to establish credible doctrine. However, the context in which data were collected is often the subject of debate as to its robustness for generalisation.

CHAOS, STRESS AND COMBAT TRAINING

War's enduring features are the actions of the enemy, climate, weather and terrain, which generate danger, friction, chaos and uncertainty.[27] The last four variables are the result of war, and are perceived by the participants. The question arises therefore as to whether one can generate training conditions that will produce perceived danger, friction, chaos and uncertainty, and allow a peacetime army to perform better in war. During combat training, leaders can be fatigued; there can be fear in company teams; reports can be inaccurate or late; mistakes can be made; activities take longer than anticipated. Such factors, combined with the subtleties of terrain and an enemy that does his best to hide and deceive, can cause significant chaos sufficient to overwhelm strong leaders.[28]

Shaw noted that, during wars:

- The number of psychiatric casualties is directly related to the intensity of combat.
- Group cohesion and expectation are powerful variables shaping response to stress.
- The individual's definition of the situation is paramount.
- All soldiers experience stress with varying degrees of mastery.
- All soldiers have a breaking point beyond which their effective performance in combat diminishes.[29]

From a research and engineering perspective it is necessary to determine what aspects of the military process are affected most by these reactions, and to design training programs that develop processes resilient to them.

Endurance and fatigue

Reporting a study by Swank and Marchand, Shaw noted that after about 30 days of combat there was noticeable decline in combat performance. Appel and Beebe found that 100 days of intermittent exposure to combat was the average length of time before non-effective behaviour became frequent.[30] Shaw also points out that the intensity of the battle (rather than its duration) was a more important variable in the cause of battle stress reactions.

Haslam et al. conducted an experiment in a tactical defensive exercise on three platoons from the parachute regiment.[31] After initial baseline testing, one platoon was scheduled no sleep, the second 1.5 hours' sleep and the third platoon three hours' sleep in 24 hours. The aim was to see how many days of total sleep loss a platoon of soldiers could endure and keep fighting effectively. The results were compared with data from platoons with partial sleep regimens. Military activities were performed and assessed, and in each 24-hour period soldiers completed a set of cognitive tests.

The findings indicated that cognitive and vigilance tasks deteriorated in proportion to the amount of sleep lost. Well-learnt physical tasks did not. Of the no-sleep platoon, 100 per cent had withdrawn after four days. On day three, the no-sleep platoon was rated as effective on physical tasks only. With increasing sleep deprivation, some junior NCOs ceased to function as leaders.

An examination of continuous operations by Frank and co-workers indicated that sustained operations produce performance decrements within 24 hours.[32] The most vulnerable tasks were those that were either long and boring or cognitively complex, with a high workload requirement. However, the effects of fatigue on sustained or continuous operations may be mitigated by working in teams.[33]

Most research studies are limited to approximately a week. The general consensus is that sleep has to be managed, and if managed well a team can endure for long periods on four hours' sleep per night. Being able to manage sleep is a combat team responsibility and something that needs to be rehearsed.

Working in a psychologically stressful environment

Environments that result in perceptions of danger, threat or lack of control among those that work in them are likely to have serious repercussions for operational performance. Although few experimental studies exist on this topic, the extant studies suggest that performance decrement depends on how strongly participants believe in the threat of death or injury.[34] Specifically, in addition to subjective anxiety and consequent negative effects on morale, experience of loss of control and fear can lead to performance 'freezing', activity and even memory at less than optimal performance levels.[35] However, this response is not universal. Experienced soldiers are able to perform their work tasks at a level comparable to the unstressed. Thus research indicates that perceptions of control (or loss thereof) are the critical precursors to performance decrements arising from fear.[36] It is also interesting to note that, as with fatigue, combat-induced stress can be moderated by group cohesion, which appears to act as a buffer against stress and boosts morale and wellbeing.[37]

Impact of chaos on teams

Most military operations are critically dependent on team performance. Although it is undoubtedly the case that the individual stressors will influence team performance on the battlefield, additional stressors (and stress moderators) will affect performance. In reviewing the impact of these variables on aircrew performance, Kanki identified three classes of stressors that affect performance: team, organisational and environmental.[38] She suggests that the team members and their relationship to each other and the leader can be critical in determining the group's ability to deal with stress; but these can also be stress-producing. Furthermore, working in environments where team members may come and go, either through rotation or casualties, may impede optimal group performance.

Skill decay is a critical issue for high-readiness troops, though almost no objective data exist on rates of decay for critical tasks. Deitchman reports that the US Army Training and Doctrine Command estimated that a tank company proficient enough to defeat a Soviet regimental attack would lose 25 per cent of its capability in three months through decaying skills and crew transfers.[39]

A constraint to be considered in devising training programs for teams using resource management training approaches is the likely longevity of the team structure. If stress and performance management in the battle setting are critically dependent on team mechanisms associated with support and communication, a breakdown in the team may strip the remaining members of their skill set. Training must ensure that individual resilience is not traded for team resilience.

Examination of army units suggests that group cohesion can serve to enhance performance when the goals of the primary group are consistent with organisational requirements, but may impede performance when they diverge. Moreover, Kellett has pointed out that in 'total wars a rapid expansion of manpower, limited training resources, and replacement problems make it more difficult to indoctrinate men into unit values and to maintain cohesion at the battalion level'.[40] Thus stressors may see the group break down more quickly in these circumstances, especially where casualties within the group are high. For example, survey data from World War II indicated that men whose companies suffered heavy casualties or witnessed atrocities reported many more fear symptoms than soldiers who had not been subject to such stresses.[41]

Role clarity is another factor critical to stress in the team environment. Once again, this can be directly linked to the issue of control. Kanki argues that effective leaders help their teams deal with the stress of a chaotic work environment by setting expectations for performance and acting in a consistent manner. Organisational responses to environmental and psychological stressors may serve to minimise debilitating performance effects in the team (and individual) contexts.[42]

MODERATING STRESS IN THE BATTLEFIELD

Prior exposure to a stressor has been shown to moderate the influence of the stressor. Norris and Murrell have argued that this is because previous experience raises the threshold at which the effects of stress become deleterious.[43] Empirical research looking at the beneficial effects of prior exposure to a stressor has validated early qualitative evidence suggesting that 'battle inoculation' was critical to the successful preparation of soldiers for combat.[44] Subsequent work with stress inoculation training, while generally focused on specific stressors, has served to demonstrate its efficacy as an approach to training as well as a clinical intervention.[45]

Bowers and co-workers have identified a number of group-level moderators of stress.[46] Variables including social support and group cohesion can act as buffers against stress. Consistent with this observation, recent attempts at training for working in a stressful environment have focused on crew resource management training and how this can assist in raising performance levels.[47]

The identification of a significant set of stress performance moderator variables is an important indicator that the stress associated with combat environments is amenable to amelioration at both individual and group levels. Although selection based on identifying individual performance on moderator variables is one approach, research suggests that training interventions designed to decrease anxiety and increase perceptions of self-control should also serve to enhance performance on the battlefield.

Training for effective performance in the combat environment

Orasnu and Backer have compared three different approaches to training and the capacity of each to improve performance under stress. Each of these approaches is based on a different premise.[48]

The first, stress training, focuses on the need to reduce the anxiety associated with stress by teaching appropriate stress management skills. This approach includes, but is not restricted to, the stress inoculation technique (SIT), which is generally flexible enough to include all other techniques and is focused on the needs of those receiving training.[49] Stress training techniques on work task performance does suggest a positive effect. A meta-analysis by Saunders et al. indicated an effect size for SIT and performance of .296.[50] Using Cohen's benchmarks for effect sizes, Saunders et al. described this as 'medium', and argued that 'stress inoculation training exerts a positive effect on enhancing performance comparable to, or stronger than, other well-established training interventions'.[51]

The second set of techniques described by Orasnu and Backer as 'skill training' focuses on increasing the durability or automaticity of the skill itself, and is premised on the notion that skills should be taught and practised for extensive periods of time under 'high-fidelity' battlefield conditions.[52] The theoretical basis for this approach is overlearning.[53] The argument is that well-rehearsed tasks become automatic, lessening the requirement for attention, and that well-drilled tasks enhance predictability and control.[54] The amount of overlearning required is task-dependent, with Driskell et al. reporting in a meta-analysis review of training that both 100 per cent and 150 per cent overlearning produced moderate to strong effects on performance.[55] However, the benefits of overlearning are quickly lost for cognitive tasks, with the performance benefit reduced by one-half after 19 days.

The third set of techniques described by Orasnu and Backer, crew resource management training (CRMT), focuses on training teams to deal with stressful situations. This technique—which was originally developed for use with air transport crews in high-risk, high-stress conditions—aims to enhance team performance through communication and defuse stress by training individuals to recognise stressful situations and stressed team members.[56] Analysis of performance improvements generated using this technique indicates that performance is improved and stress reduced, although the latter is a secondary outcome rather than the primary focus. The applicability of this result to other group settings remains an issue to be empirically addressed.

Skill training also serves to influence the attitudes of individuals facing stressful situations by increasing self-efficacy and control, thereby decreasing anxiety. Skill training itself can be viewed as a form of stress management.

Friedland and Kienan have addressed the question of how a training package, using any combination of these approaches, should be structured. Three requirements are described as essential for the efficient training of performance in stressful situations:

(a) Trainees should be given the opportunity to become familiar with the stressors characteristic of the criterion situation; such stressors should be introduced into the training process in a manner that (b) prevents the buildup of anxiety and (c) minimizes interference with the acquisition of skills that the training is designed to promote.[57]

Taking these requirements into consideration, three approaches to training were examined. The first involved trainees being exposed to stressors that gradually intensified (graduated-intensity training). The second, described as phased training, separated skill acquisition from stressor familiarisation by implementing them as separate phases of training (i.e. exposing the trainee to the stressor after the skill was acquired without the requirement to exhibit the skill). The third was described as a 'customised' approach, in which the intensity of exposure to stressors was adjusted according to individual and personality attributes.

On the basis of results from a number of laboratory and field studies, the first of which most commonly involved a visual search task (the 'work' task) during which electric shocks were administered at varying intensities (the 'stressor'), Friedland and Kienan concluded that phased training best met the requirements for effective training.[58] It ensured that skills were learnt, provided the opportunity for trainees to become acquainted with the criterion stressors, and allowed the latter to happen without sacrificing the former. They also suggested that customised training is a useful approach for small, highly specialised teams but is prohibitively costly and impractical for large units. The utility of graduated-intensity training was questioned, with results suggesting that this approach impaired training effectiveness. However, it is possible that the combination of graduated-intensity and phased training, in which trainees are gradually desensitised to the criterion stressors following overlearning on the work task, might offer the opportunity of optimising performance outcomes. Mitigating this is the difficulty of quantifying, calibrating and gradually intensifying the many and varied sources of combat stress. It is important to note Friedland and Keinan's final comment:

> Our results indicate that training under constant, high-intensity stressors and training that involves no stress are likely to prove counterproductive: The former interferes with the acquisition of skills that are relevant for effective task performance, and the latter does not sufficiently acquaint the trainee with criterion stressors.[59]

CONSTRAINTS ON COMBAT PREPARATION

A number of important constraints arise when designing and implementing training for combat. The definition and design of the training environment and the necessary level of fidelity are issues that require careful analysis. In general terms, should training packages describe what stressors should be introduced and in what form? Is it enough simply to use a stressor that reproduces similar psychological sequelae to those experienced in war (e.g. dread, fear, anxiety), through whatever means; or do the stimuli themselves have to conform to those experienced in the war? If so, how do we accurately reproduce and calibrate these stressors and ensure that individuals are not exposed to a level of acute stress that is beyond their ability to cope? Further research is needed in relation to the level of fidelity in the training environment. Certainly, research in social psychology suggests that simulated environments can reproduce high-stress psychological reactions that mimic those likely to be experienced in real environments, and that this fact is not necessarily mitigated by the realisation that the environment is only a simulation and short-lived.[60]

An ancillary issue in discussing transfer of training is the question of whether battlefield preparation is tied to specific environments and stressors. Keinan and Friedland have raised the question of whether the effect of training under one set of stressors (environments or tasks) generalises to performance with a different stressor (environment or task). If some level of generalisation is not achievable, the implications for training costs and practicality are formidable.[61]

Trainers considering the possibility of using some form of stress inoculation have to be cognisant of individual differences in resilience and vulnerability to trauma. Research suggests that experience of a traumatic event does not produce post-traumatic stress disorder (PTSD) uniformly, suggesting that individual difference variables (e.g. aspects of personality) moderate reactions to the same event.[62]

ETHICAL CONSIDERATIONS

With the likely negative impact of stress on the mental and physical wellbeing of soldiers, can we afford to realistically recreate the battlefield during training? It is impossible to simulate physical fear (fear of dying) in a training environment. Soldiers do not expect to die during a training exercise. It has been observed that, during training at a CTC, soldiers will often attempt heroic acts that they might not attempt in war.

While it is not practically or ethically reasonable to instil fear of physical harm during training, fear of failure can be manipulated. CTC training sets soldiers against an aggressive professional OPFOR (opposing force) in a 24-hour tactical battlefield. Many aspects of performance are measured via instrumentation systems (e.g. video, voice logging, GPS tracking, engagement ratios) and subject matter expert ratings. These measures give an objective evaluation of performance. Commanders are motivated to achieve their mission, outwit the enemy, and not lose assets (including lives) in the process. They are judged by their peers during the after-action review according to how well they achieved these goals. Soldiers are motivated to stay alive, as casualties are removed from the exercise. Fear of poor performance creates anxiety and stress.[63]

It is possible to create stress in the training environment by making field exercises long enough so that the effects of food and sleep deprivation become apparent. Other factors that contribute to stress are the tempo at which exercises are conducted (reducing the time available for planning and decision making) and uncertainty (perhaps as a result of deteriorated communications or poor intelligence picture).

While the provision of realistic training has ethical implications for the wellbeing of soldiers during peacetime, it has been argued that it is ethically wrong *not* to provide soldiers with a training experience likely to enhance their survivability during war. Based on the assumption that realistic combat training increases the survivability of troops in battle, there is a moral obligation to prepare soldiers for the rigours of war.

The army has a duty of care to deploy only soldiers that have reached an appropriate standard. The US Army 48th Armoured Brigade (National Guard) was trained at a CTC during the Gulf War. As they failed to reach a required standard, a decision was made not to deploy the unit.[64] CTCs can provide a means for objectively measuring training readiness.

SIMULATION AND COMBAT TRAINING

The development of a tactical digital intranet (digitisation) and the use of models and simulation have potential for improving commanders' and combat teams' performance. However, the need is to develop methods for assessing and diagnosing the skill sets required for war. A sophisticated measurement and analysis system will be required to assess skill levels of individuals and teams. Hallmark and Crowley demonstrated the usefulness of a well-constructed experiment in pointing out areas of weakness.[65] Lucas and co-workers indicate that, to improve the value of war-fighting experiments, one needs to design good experiments rather than focus on data gathering.[66]

Hallmark and Crowley indicated that some skills, identified as lacking at the NTC, could be improved by tailored training using simulation exercises at home station, and the use of JANUS was mentioned. The degree to which JANUS represents realistic human activities that can characterise chaotic conditions in war needs to be evaluated. Rooney et al. analysed simulation trials (SIMNET) to determine the impact of a digital tactical intranet on armour squadron performance.[67] One of their observations was that the simulated opposing force kept on fighting until it was eventually destroyed: '. . . computer generated forces do not, as yet, suffer from surprise and shock'. Rowland noted in urban fighting trials that all defenders kept on fighting until killed, whereas in war, prisoners, wounded, escapees and withdrawals made up 80 per cent of all defenders.[68] The outcome in war places a very different demand on attackers from training. There is an urgent need to make simulations realistic by improving the fidelity of the human behaviour they represent.[69]

The future digitised battlefield promises greater team awareness of the situation, faster reaction times and an advantage over the enemy.[70] Training the commander not to be overwhelmed by the extra information available and not to micromanage could well be a key focus for future combat training. An analysis of the training needs of commanders to make them more effective planners, decision makers and leaders should be undertaken.

Over the past five years land forces have emphasised the need to be able to conduct operations other than war and to be prepared to police peace against forces that are not conventional armies. Dealing with operations other than war may require sets of skills and NDM templates that are different from the ones traditionally trained for. A training-needs analysis of the skill sets for asynchronous warfare should be conducted.

CONCLUSION

The communication of commanders and the actions of combat teams potentially provide a rich event data protocol that is amenable to close analysis using some of the theoretical constructs of SA and NDM. Current analysis and after-action reviews are carried out by subject matter experts. There is the potential to improve the objective diagnostic information from combat training by using micro-instrumentation and online protocol analysis. Conducting experiments on combat team performance will lead to insight into deficiencies in skills, skill acquisition and retention. Experiments conducted in field settings will provide diagnostic information on the state of team and individual skill repertoires. Combat teams should be subjected to the stress of the TES environment after they have acquired and rehearsed collective processes under more benign conditions. Diagnostic information from combat-training performance results will provide direction for focused programs involving part task simulations and part home-based training.

Conclusion: the eternal human face of warfare

Michael Evans

Let us then study man in battle, for it is he who really fights.

Ardant du Picq, *Battle Studies*

GERALD F. LINDERMAN has written that 'every war begins as one war and becomes two, that watched by civilians and that fought by soldiers'.[1] The experience of the home front is so different from that of the front line that a divergence of outlook regarding the character of the fighting soon develops. In peacetime, when the civilian and military spheres reunite, the experience of war again becomes one in the public's imagination—a process that often marginalises the veteran. Combat veterans soon discover that civilian society has no place for the special code and rituals of the front line. As the war correspondent Ernie Pyle once put it, to convey war realistically means thinking about the horror of combat over and over again until 'at last the enormity of all those newly dead [strikes] like a living nightmare'.[2]

Few societies and even fewer combat veterans can contemplate an existence as a 'living nightmare'. So it is that, in a strange symbiosis, the behaviour of both civilians and veterans often works to suppress the reality of the human face of warfare. For the collective good, civilian society purifies the memory of war until, as the American poet Walt Whitman wrote, the act of fighting resembles little more than a 'quadrille in a ballroom'.[3] For their

198

part, many combat veterans find consolation in suppressing their disturbing memories by retreating into a silence that is often mistaken for a form of heroic modesty.

The chapters in this book have sought to avoid portraying the human face of war as a 'quadrille in a ballroom'. From the perspective of many disciplines, each contribution in its own unique manner has sought to explore the interior landscape of what it is like for human beings to experience warfare. As Roger Spiller points out in his introduction, there has long been a tension between the materialist and human conceptions of war. Today a materialist, instrumental conception of war dominates Western ideas about warfare. This technocentric view has arguably reached its apogee at the end of the 20th century through the notion that the West is now in the midst of a revolution in military affairs (RMA). To its most ardent proponents, the RMA will introduce information-age systems and remotely piloted weapons that will replace close combat with stand-off warfare and long-range precision strike. Yet it was Ardant du Picq that, after analysing the impact of new rapid-firing and long-range weapons in the 1860s, concluded, 'combat requires today, in order to give the best results, a moral cohesion, a unity more binding than at any other time'.[4] Du Picq's timeless message is in danger of being forgotten. As Professor Spiller points out, if direct human action is not quite obsolete, it is certainly obsolescing. It is not that the RMA should be disregarded; indeed it should be embraced, but always with the caveat that war remains a fundamentally human act. Spiller's reminder that the ADF's mission in East Timor is essentially the resolution of a sociopolitical problem is a timely reminder of the eternal human dimension of warfare.

In Chapter 1, on the psychology and physiology of combat, Dave Grossman suggests that much remains to be learnt about the behaviour of men in battle. Lieutenant Colonel Grossman provides an analysis of a scientific approach towards understanding human aggression, which he terms 'killology'. He details the factors at work in combat, including fear, aggression, tunnel vision, auditory exclusion and loss of motor control. He goes on to argue that exposure to prolonged modern combat nearly always tends to lead to some type of psychiatric disorder. Few would disagree with this proposition. However, Grossman's view—that combat studies research from Du Picq to S.L.A. Marshall through to experiments conducted by law enforcement agencies suggest 'that man is not, by nature, a killer'—is more contentious. Joanna Bourke's recent book, *An Intimate History of Killing*, argues an entirely opposite view. Bourke opens her book with a firm, even dogmatic statement: 'the characteristic act of men at war is not dying, it is killing'.[5] Yet Grossman is surely right when he argues that modern society has an inability to confront the truth about the reality of fighting and killing in war; it can only stand to gain by increasing its knowledge of both phenomena.

Colonel Steven Tetlow's chapter, on the incorporation of human factors in the British Army's work on combat simulation, employs the techniques of operational analysis. This work has been undertaken by the British Centre for Defence Analysis (CDA) and has attempted to investigate the modelling of fear and surprise as speculative research projects. The CDA's techniques are heavily dependent on quantitative research and statistical analysis. Tetlow points out that this type of analysis has assisted British Army force development by establishing a correlation between weapon numbers and weapon manning. More contentious, however, is the CDA's claim that its research has discovered a firm correlation between nationality and military effectiveness. It is perhaps unfortunate that the CDA does not identify the nationalities studied. Overall, much of the CDA's research would seem to reinforce what is already known about the human factor in combat—for example, the need for cohesive subunits. As Tetlow admits, conflict remains an intensely human activity, and the military needs to ensure that technology becomes the servant and not the master in the battle-space of the new century.

In Chapter 3 Michael Evans offers a comparison of the careers of two of the greatest combat infantrymen of the 20th century, Albert Jacka of Australia and Audie Murphy of the USA. Dr Evans' analysis employs a comparative historical method, informed by insight from both psychology and literature. He lists the remarkable similarities between the two men in terms of social background. He goes on to argue that the charismatic Jacka was probably a soldier–adventurer, who loved what the writer Tim O'Brien has called 'the awful majesty of combat'.[6] The quiet and withdrawn Murphy was, on the other hand, probably a soldier–killer, whose deadly efficiency as an infantryman was unsurpassed. Evans shows how neither man adjusted well to peacetime life: Jacka suffered from physical ill-health as a result of his wartime exploits in the World War I trenches; Murphy was clearly an undiagnosed victim of World War II PTSD. It could be argued that the terrible price exacted from both Jacka and Murphy in war is an argument against using infantry in close combat. Yet, as Evans points out, other wartime heroes underwent similar experiences and slipped back quietly into civilian society. Ultimately, Evans concludes that infantry will continue to remain the most trusted of military systems, as only infantry can provide the physical presence that is required for true victory.

The subject of the human cost of warfare is taken further by Hugh Smith's comprehensive chapter, which seeks to analyse the changing attitude towards military casualties in Western societies. Professor Smith links the Western public's anxiety about casualties to the pervasive global influence of the electronic media and the decline of the specialist defence reporter in favour of the human-interest journalist. He tackles, head-on, the proposition that Western democracies have lost the stomach for casualties. Focusing mainly on the US experience from World War II to Somalia, Smith points out that, while an aversion to casualties has grown in Western society, the reasons are highly complex. He suggests that the casualty phenomenon is partly connected to perceptions of vital interests and to other more pragmatic calculations, such as the likelihood of military success and the duration of an operation. In wars for national defence, American tolerance of casualties is likely to be high; where the cause is less vital, tolerance declines markedly. A contested cause may be aggravated by political division in Western democracies and by media attention in the form of the so-called 'CNN factor'. Aversion to casualties is also affected by sociological changes in the Western military and the phenomenon of the 'safe military'—in which more deaths are caused by accidents than by hostile action. One of Smith's most revealing pieces of evidence is that it is now more dangerous to be a journalist or UN civilian worker in a combat zone than to be a soldier or peacekeeper. The post-heroic era that has followed the end of the Cold War has not been conducive to great military crusades against regional warlords and terrorists.

Peter Warfe's chapter, on the problem of post-traumatic stress among peacekeeping forces, contains a chilling description of the Kibeho massacre in Rwanda in April 1995. At Kibeho camp, members of the Australian Medical Force serving with UNAMIR were compelled to watch the ethnic massacre of some 4000 Hutu refugees by the Tutsi-dominated RPA. The terrible paradox of being an armed peacekeeper but being unable to prevent the killing of unarmed civilians informs Colonel Warfe's chapter. There were no missiles at Kibeho—only medics supported by two Australian infantry sections. A Tutsi militia force of two battalions armed with Kalashnikovs, rifle-propelled grenades and machetes confronted the Australians. Warfe provides a powerful reminder that peace operations are still the most realistic experience of near-combat situations and may expose military personnel to extreme levels of shock and horror. As a result of the horrifying sights and sounds of Kibeho, 20 Australian service personnel have required treatment for PTSD. Colonel Warfe struck a powerful chord at the conference when he said that its success would be measured by how well Australia cares for its Rwandan veterans in future. His outline of the collaborative measures the ADF is implementing to counter the incidence of acute stress reaction and PTSD is informative. As the ADF takes up a major and long-term operational commitment to the stability of East Timor, his call for the creation of a centralised human-sciences agency may become a priority in the Department of Defence.

The dilemma of balancing deep and close battle in the future has become a major issue in Western armies in the wake of the 1990–91 Gulf War. Yet, as John English wisely reminds us in Chapter 6, much in warfare remains the same. He quotes Clausewitz's warning that war is a 'true chameleon', whose shape may change but whose nature is eternal. Future battle, Professor English argues, can be understood only through the prism of the past. He points out that the information-age concept of near-perfect situational awareness in which land forces support precision air power can be traced back to the Great War of 1914–18 and the Pentomic era of the US Army in the 1950s. In both cases artillery and nuclear strikes were declared to be more important than infantry, and both propositions were wrong. Similarly, it is premature to declare the era of inter-state conventional war to be at an end: big armies may still matter.

According to English, the Gulf War has few lessons to offer from the perspective of close combat. The disparity between the two sides was too great. For this reason English is inclined to the view that the 'non-war' of NATO Central Front of the Cold War era is of more relevance as a guide to understanding future warfare. The NATO–Warsaw Pact confrontation in Europe pitted two roughly equal and symmetrical opponents with all the traditional challenges of military science—including both deep and close battle. Such concepts as the AirLand Battle and the follow-on forces attack (FOFA) were developed by NATO, which was condemned to defend a shallow front. Significantly, AirLand Battle—a concept for war in Europe—was vindicated in the deserts of the Gulf in a conventional war no-one could have predicted in the 1980s. The message is clear: infantry divisions are not obsolete; nothing in warfare is predictable; and we ignore the lessons of NATO and the Warsaw Pact at our peril.

Like John English, Paddy Griffith in his chapter on fighting spirit on the future battlefield seeks to exploit the knowledge of the past. Dr Griffith argues that, while the battlespace of the future will be much more visible and lethal than the 'empty battlefield' of the past, the nature of warfare will not change. The age-old 'buddy group'—familiar from the time of Thucydides—will remain the key formation in combat. However, the responsibility of small units is likely to grow commensurate with increases in information-age communications and firepower. The day when a corporal can fire a cruise missile is, according to Griffith, not far away. It is likely that battle will become more difficult because of smart weapons, chemical agents and lasers, and yet safer because of precision targeting and efficient casualty evacuation. Echoing Hugh Smith, Griffith argues that in the West death while serving in the army is no longer normal, or even acceptable. He suggests, perhaps contro-versially, that during the Gulf War there were few real terrors for Western soldiers beyond the possibility of street fighting in Baghdad—something that was politically unacceptable at the time. He goes on to warn that, in both urban and jungle warfare, Western advantages in technology can be more easily nullified, putting a premium on fighting spirit, leadership and morale in military operations. It is these timeless human qualities that count in the end. Griffith believes that the Anzac storm troops on the Western Front during the Great War remain a model of what is needed to create fighting infantry in the future—infantry bonded by mate-ship, good leadership, physical toughness and efficient training.

Jim Wallace brings the perspective of a senior professional soldier to bear in his thought-provoking chapter on the interface between information-age digitisation and traditional command methods. As a practitioner, Brigadier Wallace effectively demystifies digitisation—a key feature of the RMA—and the main process by which a battlefield can be networked. Wallace provides a fascinat-ing insight into the contradictions now confronting information-age soldiers. While digitisation promises to reduce uncertainty in the field, information overload might paradoxically slow down operational decision making. Wallace points out that improved information gathering and situational awareness are not the same as intelligence processing and situational understanding. The coming of the information-age 'silent headquarters' dominated by computer screens may seduce command staffs into treating battle command and control as little more than a Nintendo video game. The ultimate danger of digitisation seems to be its potential to create detachment between headquarters staff and front-line troops. The possibility of a return to the chateau generalship of the Great War of 1914–18 may be the unwelcome offspring of command systems controlled by computer screens. Wallace's message is clear: digitisation does not mean that the ghost of Jomini armed with information-age systems has triumphed over friction and chaos in the field. War remains an art; even in the information age, battle remains the supreme contest of human wills.

Brigadier Wallace's assessment of the challenges facing tactical commanders is complemented by David Horner's insightful discussion in Chapter 9 of the stresses at work on higher commanders in future warfare. Professor Horner concentrates on both the operational and strategic levels of war. Using examples from the American Civil War, the two world wars, the Vietnam War, the Gulf War and Somalia, Horner identifies a number of enduring factors that cause stress in commanders. These factors include enemy activity, resource and physical limitations, political pressure, problems of media control and public support, cooperation with allies and human pride. Horner points out that, for all the technology available to General Schwarzkopf in 1991, he faced exactly the same question that confronted General Eisenhower on the eve of Normandy in 1944, namely the personal decision to launch a campaign. In the future, higher commanders will face ambiguous missions, the need to minimise casualties and the challenge of relying on information technology. Resolution of these challenges will require intuition, character and judgement rather than computer analysis. Like Wallace, Horner doubts that information-age technology will reduce stress on commanders. Nor does he believe that political and ethical pressures will ease off in an age of ever-increasing direct communications and unprecedented media coverage of military events. In short, the personal and essentially human challenge of command will remain as complex in the next century as it was in the past.

In 1993 John Keegan wrote, 'warfare is . . . the one human activity from which women, with the most insignificant exceptions have always and everywhere stood apart . . . Women do not fight . . . and they never, in any military sense, fight men'.[7] Warfare is 'an entirely masculine activity'.[8] Nonetheless, the issue of women in combat is now one of the most divisive issues in advanced Western militaries. The noted American military sociologist Charles Moskos opposes women in combat on the basis of biological difference, arguing that 'women are not little men and men are not just big women'.[9] A contrary view has come from Brigadier General Thomas Draude of the US Marine Corps, who argues 'I believe we must fill our ranks with our best, regardless of gender'.[10] Eleanor Hancock explores the continuing debate over women in combat in her chapter on the implications of attempting to create gender-neutral armed forces. Dr Hancock highlights how Western policies towards women in the military vary widely. Italy has no women soldiers; Germany only a handful and in restricted roles; Australia, Britain, France and the USA have enhanced the role of women in their armed forces but all retain various restrictions on combat roles; Canada, Belgium, the Netherlands and the Scandinavian countries have few restrictions or none at all.

Hancock traces the steady extension of women in combat roles to such influences as a declining birthrate, the end of conscription and the rise of modern feminism. Using the Soviet case from World War II, she argues that gender integration can be made to work and that such objections as a woman's comparative lack of upper- body strength are exaggerated and can be overcome by imaginative training regimens. Hancock tackles the difficult questions: will women kill, and will men kill women?. Too often the central question is avoided in the heat of the public policy debate. Using evidence from the Soviet armed forces in World War II and the phenomenon of femicide in Western civil society, she concludes that both questions will be answered in the affirmative. Hence, gender integration of the armed forces may not radically transform the battlefield. Hancock may be right. Unhappily, we will not know this for certain until gender-integrated units are committed to the direct-fire battlefield.

The developments in land force combat team training in recent years are analysed in Chapter 11 by Jeremy Manton, Carlene Wilson and Helen Braithwaite of the Australian Defence Science and Technology Organisation (DSTO). Like Steven Tetlow, the authors warn that there is no objective evidence that performance in peacetime land exercises can be transferred to performance in wartime. While new technologies such as tactical engagement simulation (TES) provide more realistic situations, those designing scientific combat training must strive to produce an approach that is integrated and reflects change. Echoing Jim Wallace and David Horner, Manton, Wilson and Braithwaite point out that recent research by the US Army's National Training Center demonstrates that few commanders react well to direct-fire engagements and most lack insight into the flow of battle. Skill theories, situation awareness and naturalistic decision making are among the techniques being employed to try to improve battle flow visualisation. While operational analysis can assist in preparing for combat, it cannot reproduce war's enduring features of friction, chaos and uncertainty. Scientific research into combat teams provides information that is most useful in quantifying various levels of performance in field training. The authors conclude that there is an urgent need to make simulations realistic by improving the fidelity of human behaviour. This is especially the case in the digitised battle of the future, where there is a growing need for effective measures to avoid information overload.

If any single theme runs through this book, it is the belief that technology provides no 'silver bullet' to overcome the unpredictability of the battlefield. War still resembles a thick, ghostly fog in which the human face of war remains eternally shrouded. That this judgement is reached by a diverse group of scholars, scientists and soldiers should provide a cautionary warning to those in Western defence establishments who believe that air-delivered precision-guided weapons represent the essence of future warfare. As S.L.A. Marshall once recalled, 'weapons may change . . . as they have continuously changed in the past, but the greater becomes their potential power . . . the more imperative grows the need for a greater rallying and more perfect organisation of our own human and moral forces'.[11]

In the early years of the 21st century, it is unlikely that warfare will become a bloodless, high-technology series of strikes against a static array of targets. The potential enemies of the West will surely learn from the Gulf War and from the air war over Kosovo. Future adversaries will take their military activities to the jungles and the cities where information technology is less efficient and troops on the ground are formidable opponents. To counter such forces, Western democracies will require Marshall's 'greater rallying' of human and moral forces, as relying on machines and computers will not be enough.

Endnotes

CHAPTER 1

1 This paper is built, in part, on papers previously published in the *Oxford Companion to American Military History* and the *Academic Press Encyclopedia of Violence Peace and Conflict* (Dave Grossman 1999).

2 R.A. Gabriel, *No More Heroes: Madness and Psychiatry in War,* Hill & Wang, New York, 1987.

3 R.L. Swank & W.E. Marchand, 'Combat neuroses: development of combat exhaustion', *Archives of Neurology and Psychology*, vol. 55, 1946, pp. 236–47.

4 S.L.A. Marshall, *The Soldier's Load and the Mobility of a Nation*, Marine Corps Association, Quantico, VA, 1987 (first published 1950).

5 A. Artwohl & L. Christian, *Deadly Force Encounters*, Paladin Press, Boulder, CO, 1997.

6 B.K. Siddle, *Sharpening the Warrior's Edge: The Psychology and Science of Training*, PPCT Management Systems, Millstadt, IL, 1995.

7 S.L.A. Marshall, *Men Against Fire*, Peter Smith, Gloucester, MA, 1978.

8 See in particular: P. Griffith, *Battle Tactics of the Civil War*, Yale University Press, New Haven, CT, 1989; R. Holmes, *Acts of War: The Behavior of Men in Battle*, The Free Press, New York, 1985; J. Keegan, *The Face of Battle*, The Chaucer Press, Harmondsworth, England, 1976; J. Keegan, & R. Holmes, *Soldiers*, Hamish Hamilton, London, 1985.

CHAPTER 2

1 The term 'simulation' (sometimes termed 'constructive simulation') used in this chapter is as used in the discipline of operational analysis. The author wishes to acknowledge the assistance and considerable contribution of A.J.F. Durrant Esq, Principal Operational Analyst at the Directorate of Land Warfare in the UK Ministry of Defence, in compiling this chapter, and R. Blues Esq, Technical Manager of the Historical Research Department at the UK Centre for Defence Analysis, and his team, whose vision, research and analysis are the basis of the majority of the work to which this chapter refers.

CHAPTER 3

1 Gerald F. Linderman, *The World Within War: America's Combat Experience in World War II*, The Free Press, New York, 1997, pp. 235–62.
2 S.L.A. Marshall, *Men Against Fire: The Problem of Battle Command in Future War*, Infantry Journal Press, Washington, DC, 1947, p. 26.
3 Roger J. Spiller, 'Shell shock', *American Heritage*, May/June 1990, p. 78, and 'Isen's run: human dimensions of warfare in the 20th century', *Military Review*, May 1988, LXIX, v, pp. 17–31.
4 Sir Ian Hamilton, *A Staff Officer's Scrap-Book during the Russo-Japanese War*, 2 vols, E. Arnold, London, 1906, vol. 1, p. v.
5 James Jones, *The Thin Red Line*, Charles Scribner's Sons, New York, 1951, p. 495.
6 J. Glenn Gray, *The Warriors: Reflections of Men in Battle*, Harper & Row, New York, 1959, p. 28.
7 ibid, p. 21. In Roman mythology Mars was the god of war. In Greek mythology Lethe was the river of oblivion, whose water had the power to make those who drank of it forget the whole of their former existence.
8 Roger J. Spiller, 'The price of valor', *Military History Quarterly*, Spring 1993, V, iii, p. 106.
9 Quoted in Colonel Frederick W. Timmerman, 'Human dimensions of the battlefield', *Military Review*, April 1989, LXIX, iv, p. 6.
10 Ardant du Picq, *Battle Studies: Ancient and Modern Battle*, Colonel John N. Greely & Major Robert C. Cotton (trans.), Stackpole Books, Harrisburg, PA, 1987; Marshall, *Men Against Fire*; Gray, *The Warriors*.
11 Gray, *The Warriors*, p. 40.
12 Du Picq, *Battle Studies*, ch. 2.
13 Marshall, *Men Against Fire*, pp. 89–90.
14 Gray, *The Warriors*, p. 57.
15 ibid, p. 110.

16 ibid, pp. 125–7.

17 Robin Gerster, *Big-Noting: The Heroic Theme in Australian War Writing*, Melbourne University Press, Melbourne, 1987, p. 133.

18 Quoted in Don Graham, *No Name on the Bullet: A Biography of Audie Murphy*, Viking, New York, 1989, p. 43.

19 Author's observation during a visit to Arlington, 10 October 1999. Murphy's grave—flanked by two small American flags—is situated behind the Tomb of the Unknowns.

20 Ian Grant, *Jacka, VC: Australia's Finest Fighting Soldier*, Macmillan, Melbourne, 1989, p. 28.

21 C.E.W. Bean, *The Official History of Australia in the War of 1914–18, vol. III: The AIF in France*, Angus & Robertson, Sydney, 1942, p. 720.

22 Grant, *Jacka, VC*, chs 6–9.

23 L.L. Robson interviewed in *Jacka, VC*. A film by Nigel Buesst & Ross Cooper, Monash University 1977. I am grateful to Mr Buesst for access to this important and rare film, which contains invaluable interviews with World War I veterans who served with Albert Jacka.

24 Gray, *The Warriors*, p. 110.

25 Corporal Charles Smith cited in Grant, *Jacka, VC*, p. 70.

26 ibid, pp. 97–100.

27 Quoted in Buesst & Cooper, *Jacka, VC*.

28 E.J. Rule, *Jacka's Mob*, Angus & Robertson, Sydney, 1933, p. 277.

29 Quoted in Grant, *Jacka, VC*, p. 83.

30 Rule, *Jacka's Mob*, pp. 249–53.

31 Jane Ross, *The Myth of the Digger: The Australian Soldier in Two World Wars*, Hale & Iremonger, Sydney, 1985, p. 97.

32 Gray, *The Warriors*, pp. 28–9, 51.

33 ibid, p. 28.

34 Ernie Pyle, *Here is Your War*, Lancer Books, New York, 1943, pp. 303–4.

35 E.J. Rule quoted in Grant, *Jacka, VC*, p. 84.

36 Percy Bland interviewed in Buesst & Cooper, *Jacka, VC*.

37 Grant, *Jacka, VC*, p. 125.

38 ibid, p. 160.

39 Donovan Joynt interviewed in 1977 in Buesst & Cooper, *Jacka, VC*.

40 Grant, *Jacka, VC*, p. 175.

41 See the biographies of Audie Murphy by Colonel Harold B. Simpson, *Audie Murphy: American Soldier*, Hill College Press, Hillsboro, TX, 1982, and Don Graham, *No Name on the Bullet: A Biography of Audie Murphy*, Viking, New York, 1989. For profiles of Murphy as a combat infantryman see Colonel Harold B. Simpson, 'The Audie Murphy story: combat soldier, movie actor, writer', in his *Simpson Speaks on History*, Hill College Press, Hillsboro, TX, 1986 and Roger J. Spiller,

'The price of valor', *Military History Quarterly*, Spring 1993, V, iii, pp. 100–10.

42 Graham, *No Name on the Bullet*, pp. 101–2; Spiller, 'The price of valor', pp. 100–1.

43 Quoted in the Audie Murphy Research Foundation *Newsletter*, vol. II, Spring 1997, p. 8. See also Graham, *No Name on the Bullet*, pp. 101–2.

44 Audie Murphy, *To Hell and Back*, 8th edn, Holt, Rinehart & Winston, New York, 1971.

45 ibid, pp. 207–8.

46 Graham, *No Name on the Bullet*, p. 90.

47 ibid, p. 74.

48 Tom Ryall, 'Audie Murphy' in Edward Buscombe (ed.), *The BFI Companion to the Western*, André Deutsch, London, 1988, pp. 372–3. Emphasis in original.

49 Roger J. Spiller, 'War in the dark', *American Heritage*, February/March 1999, p. 51.

50 Simpson, 'The Audie Murphy story: combat soldier, movie actor, writer', p. 9.

51 Ron Kovic, *Born on the Fourth of July*, McGraw-Hill, New York, 1976, p. 43.

52 Quoted by Graham, *No Name on the Bullet*, p. 247.

53 ibid, pp. 10–12.

54 ibid,. p. 99.

55 Roger J. Spiller, 'The tenth imperative', *Military Review*, April 1989, LXIX, iv, p. 12.

56 Garry Wills, *John Wayne: The Politics of Celebrity*, Faber & Faber, London, 1997, p. 113.

57 See Graham, *No Name on the Bullet*, ch. 14.

58 Gray, *The Warriors*, p. 57.

59 Murphy, *To Hell and Back*, p. 11.

60 ibid, p. 125.

61 Quoted in Graham, *No Name on the Bullet*, p. 274.

62 Kirk Douglas, *The Ragman's Son: An Autobiography*, Pan Books, London, 1988, p. 182.

63 Quoted in Spiller, 'The price of valor', p. 108; Graham, *No Name on the Bullet*, pp. 319–20.

64 Interview with Denver Pyle, Audie Murphy Research Foundation *Newsletter*, vol. III, Winter 1998, pp. 1–2. A similar view of Murphy was held by the Hollywood director, Bud Boetticher. See Graham, *No Name on the Bullet,* p. 274.

65 Graham, *No Name on the Bullet*, pp. 319–20.

66 This theme is well analysed in Paul Zweig, *The Adventurer: The Fate of Adventure in the Western World*, Princeton University Press, Princeton, NJ, 1974.

67 ibid, p. 223.

68 Marcus Aurelius, *Meditations*, Wordsworth Classics, Hertfordshire, 1997 edn, book 4, p. 33.

69 Merrilyn Lincoln, 'Henry William Murray', in Chris Coulthard-Clark (ed.), *The Diggers: Makers of the Australian Military Tradition*, Melbourne University Press, Melbourne, 1993, pp. 270–2.

70 Like Murphy, Britt was a Southern farm boy and belonged to the Third Infantry Division. For Britt's background see Graham, *No Name on the Bullet*, pp. 118–19.

71 Zweig, *The Adventurer*, p. 223.

72 Commonwealth of Australia, *Parliamentary Debates*, House of Representatives, vol. 207, 25 May 1950, p. 3211.

73 ibid, p. 3212.

74 ibid.

75 Henry 'Jo' Gullett, *Good Company*, University of Queensland Press, Brisbane, 1992, p. 250. Gullett was the member for Henty 1946–55. His memoir, *Not As A Duty Only: An Infantryman's War*, Melbourne University Press, Carlton, 1976, is one of the finest to emerge from World War II.

76 *Jacka, VC*, film by Buesst & Cooper.

77 Grant, *Jacka, VC*. See also K.S. Inglis, 'Anzac and the Australian military tradition', *Revue Internationale d'Histoire Militaire*, Canberra, no. 72, 1990, pp. 1–24.

78 *Time Australia*, 25 October 1999, pp. 50–1.

79 ibid. The list was C.E.W. Bean, Albert Jacka, John Monash, John Simpson Kirkpatrick, Arthur Roden Cutler, Nancy Wake, the 'Fuzzy Wuzzy Angels' (the Papuans and New Guineans of the Kokoda Trail), Ernest Edward 'Weary' Dunlop and Charles Upham.

80 Sue Gossett, *The Films and Career of Audie Murphy*, Empire Publishing, Madison, NC, 1996. For the international interest in Murphy see the Audie Murphy Research Foundation *Newsletter*, vol. I, Winter 1997, pp. 1–3.

81 Audie Murphy Research Foundation Secretary/Treasurer's Report, Winter 1997 in the Foundation's *Newsletter*, vol. 1, Winter 1997.

82 See for example Audie Murphy Research Foundation *Newsletter*, vol. 5, Summer/Fall 1998 and vol. 7, Spring 1999.

83 Victor Davis Hanson & John Heath, *Who Killed Homer? The Demise of Classical Education and the Recovery of Greek Wisdom*, The Free Press, New York, 1998, p. 75.

84 See Edward N. Luttwak, 'Towards post-heroic warfare', *Foreign Affairs*, May/June 1995, LXXIV, iii, pp. 109–22, and 'A post-heroic military policy', *Foreign Affairs*, July/August 1996, LXXV, iv, pp. 33–44.

85 Roger J. Spiller, 'My guns: a memoir of the Second World War', *American Heritage*, December 1991, p. 47.

86 Maurice de Saxe, 'My Reveries upon the Art of War' in Brigadier General Thomas R. Phillips (ed.), *Roots of Strategy*, vol. 1, Stackpole Books, Harrisburg, PA, 1985, p. 189.

CHAPTER 4

1 Janice Gross Stein, 'Deterrence and compellence in the Gulf, 1990–91, *International Security*, Fall 1992, XVII, ii, p. 175.

2 David Malone, 'Haiti and the international community: a case study', *Survival*, Summer 1997, XXXIX, ii, p. 133.

3 International Institute for Strategic Studies, 'The problem of combat reluctance', *Strategic Survey*, 1995/96, Oxford University Press, Oxford, 1996, p. 54.

4 While armed with rifles and pistols for self-defence, Japanese troops were prohibited by law from defending other members of UNTAC even when they were under attack. See Trevor Findlay, *Cambodia: The Legacy and Lessons of UNTAC*, SIPRI Research Report no. 9, Oxford University Press, Oxford, 1995, p. 133.

5 For a comparison of democratic and non-democratic states with regard to casualties see T.S. Milburn, 'Casualties—the crucial factor in modern conflicts', *The British Army Review*, August 1996, no. 113, pp. 78–84.

6 International Institute for Strategic Studies, 'The problem of combat reluctance', p. 56.

7 Paul Fussell, *The Boy Scout Handbook and Other Observations*, Oxford University Press, New York, 1982, p. 234, and *Wartime*, Oxford University Press, New York, 1989, p. 269.

8 Graeme Dobell, 'The media's perspective on peacekeeping', in Hugh Smith (ed.), *Peacekeeping: Challenges for the Future*, Australian Defence Studies Centre, Canberra, 1993, p. 44.

9 Cited in Philip Knightley, *The First Casualty*, Harcourt Brace Jovanovich, New York, 1975, p. 411.

10 On the use of poll data see John E. Mueller, *War, Presidents and Public Opinion*, Wiley, New York, 1973, ch. 1; and *Policy and Opinion in the Gulf War*, Chicago University Press, Chicago, IL, 1994, ch. 1.

11 Scott Sigmund Gartner & Gary M. Segura, 'War, casualties and public opinion', *Journal of Conflict Resolution*, June 1998, XXXXII, iii, pp. 278–87.

12 Benjamin C. Schwarz, *Casualties, Public Opinion and US Military Intervention*, RAND, Santa Monica, CA, 1994, pp. 13–16.

13 Mueller, *War, Presidents and Public Opinion*, p. 60.

14 Charles A. Stevenson, 'The evolving Clinton Doctrine on the use of force', *Armed Forces & Society*, Summer 1996, XXII, iv, p. 515.

So high were the hurdles that Secretary of State George Schultz later termed them 'a counsel of inaction bordering on paralysis', loc. cit.

15 Eric V. Larson, *Casualties and Consensus: The Historical Role of Casualties in Domestic Support for US Military Operations*, Rand, Santa Monica, CA, 1996, pp. 36–9, 114–5. Larson notes that measurement of prospective support in polls approximated to observed support in the case of Vietnam and the Gulf, p. 10 n5. See also Mueller, *Policy and Opinion in the Gulf War*, tables 94–5, pp. 234–5.

16 ibid, pp. 76–7.

17 Schwarz, *Casualties, Public Opinion and US Military Intervention*, p. 26. A poll in Britain showed that 6 per cent of respondents wanted to 'nuke Baghdad'; the figure rose to 21 per cent among tabloid readers. See Martin Shaw, *Post Military Society*, Polity Press, Cambridge, 1991, p. 205.

18 International Institute for Strategic Studies, 'The problem of combat reluctance', p. 52.

19 Michael W. Alvis, *Dying for Peace: Understanding the Role of Casualties in US Peace Operations*, US Institute for Peace, Washington, DC, 1998 (microfiche), p. 8.

20 'Address to the nation on Somalia', 7 October 1993.

21 Steven Kull, 'What the public knows that Washington doesn't', *Foreign Policy*, Winter 1995/96, no. 101, p. 112.

22 ibid.

23 Malone, 'Haiti and the international community', p. 133.

24 Morris Morley & Chris McGillion, ' "Disobedient" generals and the politics of redemocratization: the Clinton administration and Haiti', *Political Science Quarterly*, Fall 1997, CXII, iii, p. 370.

25 ibid.

26 Alvis, *Dying for Peace*, pp. 14–15, 20.

27 Larson, *Casualties and Consensus*, pp. xvi–xviii, proposes a hierarchy with four categories.

28 Kull, 'What the public knows that Washington doesn't', p. 105.

29 Piers Robinson, 'The CNN effect: can the news media drive foreign policy?', *Review of International Studies*, April 1999, XXV, ii, p. 303.

30 Jonathan Mermin, 'Television news and American intervention in Somalia: the myth of a media-driven foreign policy', *Political Science Quarterly*, Fall 1997, CXII, iii, p. 386ff. Significantly, three other major channels—ABC, CBS and NBC—did not increase their level of coverage after the deaths in Mogadishu. Larson, *Casualties and Consensus*, p. 46.

31 Carlyle A. Thayer, 'Vietnam: a critical analysis', in Peter R. Young (ed.), *Defence and the Media in Time of Limited War*, Frank Cass, London, 1992.

32 Larson, *Casualties and Consensus*, pp. 71–2.

33 Edward N. Luttwak, 'Towards post-heroic warfare', *Foreign Affairs*, May/June 1995, LXXIII, iv, p. 115.
34 Francis Fukuyama, 'Women and the evolution of world politics', *Foreign Affairs*, September/October 1998, LXXVII, v, p. 38.
35 Charles C. Moskos, 'When Americans accept casualties', *Newsletter*, Inter-University Seminar on Armed Forces and Society, Winter 1996, p. 12.
36 Edward N. Luttwak, 'From Vietnam to *Desert Fox*: civil military relations in modern democracies', *Survival*, Spring 1999, XXXXI, i, p. 112, fn 7.
37 Peter G. Peterson, 'Gray dawn: the global aging crisis', *Foreign Affairs*, January/February 1999, LXXVIII, i, p. 44.
38 ibid, pp. 44–6.
39 Eliot A. Cohen, 'Tocqueville on war', *Social Philosophy & Policy*, Autumn 1985, III, i, pp. 207–11.
40 Michael Mandelbaum, 'Learning to be warless', *Survival*, Summer 1999, XXXXI, ii, p. 151.
41 Robert S. McNamara, *In Retrospect: The Tragedy and Lessons of Vietnam*, Times Books, New York, 1995.
42 Geoffrey Robertson, *The Justice Game*, Vintage, London, 1999, ch. 14.
43 John E. Mueller, 'Changing attitudes towards war: the impact of the First World War', *British Journal of Political Science*, January 1991, XXI, ii, p. 12.
44 Russell Weigley, *The Age of Battles*, Pimlico, London, 1993, pp. 537–40.
45 John Keegan, *A History of Warfare*, Hutchinson, London, 1993, p. 59.
46 Eliot A. Cohen, *Citizens and Soldiers: The Dilemmas of Military Service*, Cornell University Press, Ithaca, NY, 1985, p. 115.
47 ibid.
48 Harvey M. Sarpolsky & Jeremy Shapiro, 'Casualties, technology and America's future wars', *Parameters*, Summer 1996, XXVI, ii, p. 124.
49 Amnon Sella, *The Value of Human Life in Soviet Warfare*, Routledge, London, 1992, pp. 129–30.
50 Charles C. Moskos & John Whiteclay Chambers (eds), *The New Conscientious Objection: From Sacred to Secular Resistance*, Oxford University Press, New York, 1993, ch. 1.
51 Ian Wing, 'Selective conscientious objection and the Australian Defence Force', *Australian Defence Force Journal*, July/August 1999, no. 137.
52 Charles C. Moskos & John Sibley Butler, *All That We Can Be: Black Leadership and Racial Integration the Army Way*, Basic Books, New York, 1996, pp. 8–9. Black fatalities remained proportional in the operations after Vietnam.
53 Cynthia H. Enloe, 'The politics of constructing the American woman soldier', in Elisabetta Addis, Valeria E. Russo & Lorenza Sebesta (eds),

Women Soldiers: Images and Realities, St Martin's Press, New York, 1994, p. 101.

54 Linda Bird Francke, *Ground Zero: Gender Wars in the Military,* Simon & Schuster, New York, 1997, ch. 3.

55 Martin Binkin, *Who Will Fight the Next War?*, Brookings Institution, Washington, DC, 1993, p. 49.

56 ibid, pp. 50–1. There is an important question of definition, as military personnel (mostly male) that are responsible for child maintenance payments to a custodial parent may be counted as a single parent. Public concern focuses more on single parents in the military (mostly female) that have actual custody of children.

57 ibid, p. 50.

58 Moskos, 'When Americans accept casualties', p. 12.

59 Edward N. Luttwak, 'Where are the great powers? At home with the kids', *Foreign Affairs*, July/August 1994, LXXIII, iv, p. 25.

60 Charles J. Dunlap, Jr, *Technology and the 21st Century Battlefield: Recomplicating Moral Life for the Statesman and the Soldier*, Strategic Studies Institute, US Army War College, Carlisle, PA, 15 January 1999, p. 29.

61 Committee to Protect Journalists, *http://www.cpj.org/attacks98/10yrKilled Chart.html* (accessed 18 August 1999).

62 Associated Press, *Army*, 6 August 1998.

63 Cited in Christopher Coker, 'Post-modern war', *RUSI Journal*, June 1998, CXXXXIII, iii, p. 8.

64 Martin van Creveld, *The Transformation of War*, Free Press, New York, 1991.

65 Robert Burns, 'North Korea war plan may count on US losses', *Seattle Times*, 22 October 1997.

66 Edward N. Luttwak, 'A post-heroic military policy', *Foreign Affairs*, July/August 1996, LXXV, iii, p. 39.

67 David Tucker, 'Fighting barbarians', *Parameters*, Summer 1998, p. 70.

68 Anne Burns, 'Downer warns of Timor body bags', *West Australian*, 4 March 1999.

69 Karl W. Eikenberry, 'Take no casualties', *Parameters*, Summer 1996, XXVI, ii, p. 112.

70 D. Mercer, G. Muncham & K. Williams (eds), *The Fog of War*, Heinemann, London, 1987, pp. 6, 22.

71 For a debunking of this argument, see Michael O'Hanlon, 'Can high technology bring US troops home?', *Foreign Policy*, Winter 1998/99, no. 113.

72 Coker, 'Post-modern war', p. 14.

73 Eikenberry, 'Take no casualties', pp. 116–17.

74 Luttwak, 'A post-heroic military policy', p. 41.

75 Cathy Downes, 'An ethos for an army in the twenty-first century', in Hugh Smith (ed.), *Preparing Future Leaders*, Australian Defence Studies Centre, Canberra, 1998, p. 74.

76 Alvis, *Dying for Peace*, p. 13.
77 Kevin S. Woods, 'Limiting casualties: imperative or constraint?', School of Advanced Military Studies, US Army Command and General Staff College, 1997, pp. 29–30 (microfiche). No reference to casualties appears in the Australian Army's *Land Warfare Doctrine 1: The Fundamentals of Land Warfare*, DPS: 34399/99, Doctrine Wing, CATDC, 1999.
78 Luttwak, 'Where are the great powers? At home with the kids', p. 28.
79 Patricia Zengel, 'Assassination and the law of armed conflict', *Military Law Review*, Fall 1991, CXXXIV, pp. 144–5.
80 Luttwak, 'Towards post-heroic warfare', p. 115.

CHAPTER 6

1 Jacob Kipp, 'Two views of Warsaw: the Russian Civil War and Soviet operational art, 1920–1932', in *The Operational Art: Developments in the Theories of War*, B.J.C. McKercher & Michael Hennessy (eds), Praeger, Westport, 1996, p. 75.
2 Carl von Clausewitz, *On War*, Michael Howard & Peter Paret (ed. and trans.), Princeton University Press, Princeton, NJ, 1989, p. 89.
3 Martin van Creveld, *Technology and War: From 2000 BC to the Present*, Brasseys, London, 1991.
4 Dilip Hiro, *The Longest War: The Iran–Iraq Military Conflict*, Paladin, London, 1990, p. 1.
5 For a critique of US Army performance in the Gulf see Robert R. Leonhard, *The Art of Maneuver: Maneuver-Warfare Theory and AirLand Battle*, Presidio, Novato, CA, 1991, pp. 261–99. Most alarming was his charge that supporting artillery fire was poorly integrated at unit level and that there was a dearth of smoke and high-explosive shells for suppression (as opposed to destruction), which oversight almost cost the Allies the war in the Western Desert during World War II.
6 The classic instance still being that of the British Empire's triumphant 'Hundred Days of Victory' from the 'black day of the Germany army' on 8 August 1918 to Armistice Day on 11 November. Had the combined-arms warfighting techniques used during this period been maintained, *Blitzkrieg* would not have been so shocking to Anglo-French forces in World War II. Fixation with the trench warfare of the Somme and Passchendaele, however, eclipsed the progressive advances of the Hundred Days and lent credence to the myth that armies always unwisely prepared to fight the last war.
7 General Robert Close, *Time for Action*, Brassey's, Oxford, 1983, p. 106.
8 OBL Christian Millotat, 'The reinforced Panzergrenadier battalion in defence,' Presentation Combat Training Centre Tactics Study Session, CFB Gagetown, 5–7 March 1984.

9 Richard E. Simpkin, *Race to the Swift: Thoughts on Twenty-First Century Warfare*, Brassey's, London, 1985, pp. 83–4, 92.

10 That this reality was not necessarily recognised, however, was evident in the design of a proposed Canadian mechanised infantry division for the 1990s. With a total strength of 34 000, it mustered only 411 more infantry than the 8400 of World War II's 18 000-man infantry division, usually supported by a 3500-strong tank brigade. The extra 411 infantry thus formed but a small part of an overall 12 500 increase. See John A. English, 'New doctrines for infantry', *Tactics and Technology*, Canadian Institute of Strategic Studies, Toronto, 1986, p. 80.

11 ibid. During World War II, German divisions were considered 'used up' when their victualling strength dropped from 12 000 to 11 000; those that were 'totally battle-weary' rarely fielded less than 10 000. General Frido Senger und Etterlin, *Neither Fear nor Hope*, George Malcolm (trans.), Macdonald, London, 1960, pp. 196, 223.

12 Lieutenant-General J.A.R. Gutknecht, 'NATO responses: air/land battle 2000 and Follow-on Forces Attack', *Tactics and Technology*, Canadian Institute of Strategic Studies, Toronto, 1986, pp. 8–10; Robert A. Gessert, 'The airland battle and NATO's new doctrinal debate', *RUSI: Journal of the Royal United Services Institute for Defence Studies*, June 1984, pp. 54–5; and Stephen J. Flanagan, *NATO's Conventional Defences: Options for the Central Region*, Macmillan, London, 1988, pp. 87–8. Some also noted that it raised the risks of conventional war. General Sir Hugh Beach, 'No first use: a scenario for non-nuclear war in Europe today', Special Seminar on Strategic Issues, Operational Research and Analysis Establishment, Ottawa, June 1984; and Boyd D. Sutton, John R. Landry et al., 'New Directions in Conventional Defence?', *Survival*, March/April 1984, pp. 50–70.

13 Allied Forces Central Europe (AFCENT) together with Allied Air Forces Central Europe (AAFCE) formed a joint war headquarters, as did NORTHAG with the Second Allied Tactical Air Force (2 ATAF), and CENTAG with the Fourth Allied Tactical Air Force (4 ATAF).

14 General Bernard W. Rogers, 'Follow-on forces attack (FOFA): myths and realities', *NATO Review*, 6 December 1984, pp. 1–9; Flanagan, *NATO's Conventional Defences*, pp. 62, 83–4, 90, 100, 103; and Richard Mills, 'Follow-on forces attack: the need for a critical assessment', *Defence Yearbook 1990*, Royal United Services Institute and Brassey's, London, 1990, pp. 125–37.

15 Flanagan, *NATO's Conventional Defences*, pp. 87–9, 93, 103; and General Sir Hugh Beach, 'On improving NATO strategy', *The Conventional Defense of Europe: New Technologies and New Strategies*,

Andrew J. Pierre (ed.), Council on Foreign Relations, New York, 1986, pp. 173–5.

16 Peter Scharfman, 'The future of FOFA', in *Military Strategy in a Changing Europe*, Brian Holden Reid & Michael Dewar (eds), Brasseys, London, 1991, pp. 96–102; Flanagan, *NATO's Conventional Defences*, pp. 63–71; and Andrew J. Pierre, 'Enhancing conventional defense: a question of priorities', *Conventional Defense of Europe*, pp. 13–17. The purpose of JSTARS, which provided real-time monitoring and identification of stationary and moving ground targets at ranges over 50 kilometres or more, was to perform the same surveillance and battle management functions for ground warfare that the Airborne Warning and Control System (AWACS) and Joint Tactical Information Distribution System (JTIDS) performed for aerial combat (JTIDS linked NATO ground and air communications networks to 18 E–3A AWACS aircraft). In order to deal with Warsaw Pact second strategic echelon forces, FOFA also required new ballistic and cruise missile weapons systems.

17 Rogers originally argued that a 4 per cent annual real rise in national defence budgets would cover the cost of FOFA, but the reality was closer to 7 per cent. See Rogers, 'Follow-on forces attack (FOFA): myths and realities', pp. 1–9; Beach, 'On improving NATO strategy', pp. 169–70. No NATO nation except the US ever met this increase.

18 Scharfman, 'The future of FOFA', pp. 96, 99–104.

19 The second-echelon threat came third. Not surprisingly, the Soviets aimed to improve the air interdiction of NATO's *only* echelon. Flanagan, *NATO's Conventional Defences*, pp. 63, 88, 103–4; David Greenwood, 'The impact of resource constraints', in *Military Strategy in a Changing Europe*, pp. 69, 71–3, 77; Pierre, 'Enhancing conventional defense: a question of priorities', p. 21; Gotz Sperling, *German Perspectives on the Defence of Europe: An Analysis of Alternate Approaches to NATO Strategy*, Queen's Centre for International Relations, Kingston, 1985, p. 110; and Beach, 'On improving NATO strategy,' p. 169.

CHAPTER 7

1 My 'futuristic' books were P. Griffith & Elmar Dinter, *Not Over by Christmas*, Antony Bird, Chichester, 1983, which was overtaken by the end of the Cold War; and P. Griffith, *The Ultimate Weaponry*, Sidgwick & Jackson, London, 1991, which was overtaken by the Gulf War of 1991. Still to be overtaken, apparently, is my article on decentralised tank combat: P. Griffith, 'Countering surprise by mobility: a concept for armoured warfare on the Central Front', *The Sandhurst Journal*, vol. 1, no 1, 1990.

2 See P. Griffith, 'British armoured warfare in the Western Desert 1940–43', *Armoured Warfare*, J.P. Harris & F.N. Toase (eds), Batsford, London, 1990, pp. 70–87.

3 See Lex McAulay, *The Battle of Long Tan: 'The Legend of ANZAC Upheld'*, Arrow, London, 1987; and P. Eddy & M. Linklater (eds) *The Falklands War*, Sunday Times Insight Team and Sphere Books, London, 1982, pp. 7–22, 64–94.

4 See Peter Lewis-Young, 'Australia's Project Wundurra', *Asian Defence Journal*, no. 7, 1997, p. 20; Peter Lewis-Young, 'Australia's future infantryman', *Armed Forces Journal International*, February 1999, p. 44. I am also grateful to H.P. Wilmott for pointing out that the US Marine Corps is pursuing similar studies: see Marine Corps Combat Development Command, *Warfighting Concepts for the 21st Century*, USMC, Quantico, VA, 1998.

5 Although not strictly a feature of the battlefield's 'emptiness', it is noticeable that a decreasing observance of strict drills seems to accompany an increasing dispersion of the combatants. For some thoughts, see J. English ' "Keep step and they cannot hurt you": the value of drill in the Peninsular War', in *The Peninsular War: Aspects of the Struggle in the Iberian Peninsula*, Ian Fletcher (ed.), Spellmount, Staplehurst, Kent, 1998, pp. 163–72.

6 S.L.A. Marshall, *Men Against Fire*, Morrow, New York, 1947. Compare General von Heydte's account of German paratroopers on Crete huddling together for psychological comfort.

7 I particularly remember the emphatic use of this phrase by General F.-J. Schulze during the 1982 ESECS conference on Strengthening Conventional Deterrence in Europe (published as European Security Study, *Strengthening Conventional Deterrence in Europe: Proposals for the 1980s*, Macmillan, London 1983, see p. 148).

8 John Keegan, *The Face of Battle*, Jonathan Cape, London, 1976 (see especially the concluding chapter).

9 The life expectancy for workers in Britain was 39.91 years in 1838–54; 43.66 years in 1881–90; 44 years for men and 48 for women in 1901–22. In 1901 the death rate among working miners was 3.4 times that of all workers, and 10 times that of gardeners; while in 1911 the total death rate in industrial Manchester was double that of the more comfortable Bournemouth. In 1931, 4.11 women died in childbirth per 1000 births: in 1965 the figure was 0.25. I am very grateful to Dr F.R.G. Noon for supplying me with these statistics.

10 John Pimlott & Stephen Badsey (eds), *The Gulf War Assessed*, Arms & Armour, London, 1992.

11 The creation of tribalism in the military context is well described in Richard Holmes, *Firing Line*, Jonathan Cape, London, 1985. For its inner mechanics in a civilian urban environment see James Patrick, *A Glasgow Gang Observed*, Eyre Methuen, London, 1973.

12 Ardant du Picq, *Battle Studies*, J.N. Greely & R.C. Cotton (trans.), Macmillan, New York, 1921.
13 ibid, p. 226.
14 See Paddy Griffith, *Battle Tactics of the Western Front: The British Art of Attack, 1916–18*, Yale University Press, New Haven, CT, 1994.

CHAPTER 8

1 The reduction in logistics staff was achieved by a transfer of much of the planning responsibility to the commanding officer of the Combat Service Support (CSS) battalion. This transfer was made possible by digital links.
2 *Mission Command and Battlefield Digitisation: Human Services Considerations*, Defence Engineering and Research Agency, March 1998, para 2.27.

CHAPTER 9

1 Lieutenant General Charles G. Cooper, 'The day it became the longest war', *Proceedings of the US Naval Institute*, May 1996, pp. 77–80. This account is confirmed in Lewis Sorley, *Honorable Warrior: General Harold K. Johnson and the Ethics of Command*, University Press of Kansas, Lawrence, KS, 1998, pp. 222–3 and footnotes.
2 Sorley, *Honorable Warrior*, p. 304.
3 Tom Cox, *Stress*, Macmillan, London, 1978, p. 18. For a discussion of stress and stressors with respect to commanders see Glynis Breakwell & Keith Spacie, *Pressures Facing Commanders*, The Occasional No. 29, Strategic & Combat Studies Institute, Camberley, 1997.
4 The following factors are based on Cary L. Cooper, *The Stress Check*, Prentice-Hall, Englewood Cliffs, NJ, 1981, pp. 20–31.
5 Sir Archibald Wavell, *Generals and Generalship*, Times Publishing, London, 1941, p. 2.
6 B.L. Montgomery, *A History of Warfare*, Collins, London, 1968, p. 23.
7 Napoleon's Maxims, quoted in T.R. Phillips (ed.), *Roots of Strategy*, Military Service Publishing, Harrisburg, PA, 1943, p. 235.
8 Field Marshal Sir William Slim, 'Higher command in war', *Military Review*, May 1990, pp. 10–21.
9 J.F.C. Fuller, *Generalship, Its Diseases and Their Cure*, Faber, London, 1933, pp. 32, 87–8.
10 Colonel J.F.C. Fuller, *The Foundations of the Science of War*, Hutchinson, London, 1925, p. 216.

11 General André Beaufre, *An Introduction to Strategy*, Faber, London, 1965, p. 22.
12 William S. Lind, *Maneuver Warfare Handbook*, Westview Press, Boulder, CO, 1985, p. 5.
13 D.M. Horner, *Australian Higher Command in the Vietnam War*, Strategic and Defence Studies Centre, Canberra, 1986, p. 43.
14 David Horner, *Blamey: The Commander-in-Chief*, Allen & Unwin, Sydney, 1998, p. 265.
15 Bruce Catton, *Grant Takes Command*, J.M. Dent, London, 1970, p. 177.
16 John Connell, *Wavell: Scholar and Soldier*, Collins, London, 1964, p. 265.
17 *The Age*, 6 August 1942, p. 20.
18 John Hetherington, *Blamey: Controversial Soldier*, Australian War Memorial, Canberra, 1973, p. 240.
19 Major General Sir Kingsley Norris to author, 27 June 1974.
20 Hetherington, *Blamey*, p. 241–2.
21 J. Burns, *The Brown and Blue Diamond at War*, 2/27th Battalion Ex-Servicemen's Association, Adelaide, 1960, p. 147.
22 Rowell to Vasey, 30 August 1942, Australian War Memorial: AWM 54, 225/2/5.
23 Rowell to Vial, 28 September 1942, from Brigadier R.R. Vial.
24 John Robertson, *Anzac and Empire*, Hamlyn, Melbourne, 1990, p. 77.
25 Bob Breen, *A Little Bit of Hope, Australian Force—Somalia*, Allen & Unwin, Sydney, 1998, p. 43.
26 ibid, p. 235.
27 Ian McNeill, *To Long Tan*, Allen & Unwin, Sydney, 1993, pp. 252–5.
28 Slim, 'Higher Command in War', p. 12.
29 Quoted in Hetherington, *Blamey*, p. 194.
30 George C. Kenney, *General Kenney Reports*, Office of Air Force History Reprint, Washington, DC, 1987, p. 151.
31 General Norman H. Schwarzkopf, *It Doesn't Take a Hero*, Linda Grey Bantam Books, New York, 1992, p. 332.
32 Brett Lodge, *Lavarack: Rival General*, Allen & Unwin, Sydney, 1998, p. 106.
33 Robert L. Eichelberger, *Our Jungle Road to Tokyo*, Battery Classics, Nashville, TN, 1989, p. 21.
34 Schwarzkopf, *It Doesn't Take a Hero*, pp. 443–4.
35 Paul T. Harig, 'The digital general: reflections on leadership in the post-information age', *Parameters*, Autumn 1996, p. 133.
36 For a discussion of asymmetric wars see Lawrence Freedman, *The Revolution in Strategic Affairs*, Adelphi Paper 318, Oxford University Press, Oxford, 1998.
37 Breen, *A Little Bit of Hope*, pp. 266, 328.

38 Paul T. Harig, 'The digital general: reflections on leadership in the post-information age', *Parameters*, Autumn 1996, pp. 139–40.
39 ibid, p. 360.
40 Breakwell and Spacie, *Pressures Facing Commanders*, p. 11.
41 Glenda Y. Nogami, Christopher D. Brander, Karen A. Slusser, '21st century strategic leader competencies: from the army generals' perspective', *Proceedings of the 39th Annual Conference of the International Military Testing Association*, Sydney, 1997, p. 340.
42 B.H. Liddell Hart, *Thoughts on War*, Faber, London, 1943, p. 219.

CHAPTER 10

1 In both cases the country is Norway: 'C. Frauen in den Streitkräften anderer Staaten', 'Frauen in der Bundeswehr', *http://www.bundeswehr. de/bundeswehr/frauen/c-anderek.htm* (accessed 9 March 1999); Michael Whittaker, 'We are not alone—women in NATO', Defence 2000, *http://131.137.255.5/vcds/mccrt/cmrs/dec98/art08_e.asp* (accessed 5 March 1999).

2 'Women win fight', *The Australian*, 13 January 2000, p. 8; cross-post from H-Minerva by Sharon Halevi, discussion thread 'The Army Makes Men', H-War, *http://www.h-net.msu.edu/~war/*, (accessed 11 January 2000); Alan Hall, 'Lone woman defeats the German army', *The Times*, 12 January 2000, p. 15; 'Italy's unisex army flings fashion at front line', *The Weekend Australian*, February 12–13, 2000, p. 17.

3 This chapter is based on a paper given to the Land Warfare Studies Centre Conference, 'The Human Face of Warfare: Past, Present and Future', Canberra, 25 March 1999, and I am grateful to participants in the conference for their comments. It arises out of a wider research project on women and the state. Some earlier reflections on this issue are contained in 'Women, combat and the military', *Women and the State: Australian Perspectives*, Renate Howe (ed.) (a special edition of *Journal of Australian Studies*, no. 37, Bundoora, 1993), pp. 88–98. I am grateful to Major Jukka Haltia, Information Division, Defence Staff, Finland; Ms Ewa Walczuk, Information Officer, Finnish Embassy, Canberra; Ms Hanni M. Gredsted, Information Officer, Royal Danish Embassy, Canberra; staff in the Commonwealth Department of Defence; the Defence Departments and Embassies of Switzerland, Italy, Norway, Finland, Sweden, Germany and the USA; and the High Commissions of Canada, New Zealand and the United Kingdom for their assistance in providing information on the policies of their respective governments; Ms Julie Burbidge, whose contributions as a research assistant were invaluable; my colleague Professor David Garrioch for his translations from Swedish and for providing specific references; and my colleague Dr Claudia Prestel, for providing

specific references. Dr Prestel's student, Ms Einat Orbach, shared the findings of her current research on women in the Israeli military. I also wish to thank my former colleagues at the University College of the University of New South Wales, Professors Jeff Grey, History, and Hugh Smith, Politics, for their help in securing copies of otherwise inaccessible materials. In particular I am grateful for the helpful comments on earlier drafts of this paper by Lt Col. Catherine Lewis of the Australian Defence Forces, my colleague, Mr Tony Wood, of the Monash University History Department, Mr Edward Wilson, and my husband, Professor Gunther E. Rothenberg.

4 Dr Clare Burton, *Women in the Australian Defence Force: Two Studies*, Canberra, December 1996, ES–6. Current combat exclusions have further discriminatory effects because they drastically reduce women's chances of achieving top rank and increase the likelihood that they will retire earlier than their male counterparts: ibid, ES–17 – ES–20; Gervase Greene, 'Military women in danger of being a passing parade', *The Age*, 15 January 1999, p. 1; Gervase Greene, 'Crisis among defence force women denied', *The Age*, 21 January 1999, p. 5; Hugh Smith & Ian McAllister, 'The changing military profession: integrating women in the Australian Defence Force', *Australia and New Zealand Journal of Sociology*, 27/3, November 1991, p. 374.

5 For examples of this participation, see Julie Wheelwright, *Amazons and Military Maids: Women who Dressed as Men in the Pursuit of Life, Liberty and Happiness*, Pandora, London, 1989, pp. 39–40 (Serbia), 29–34, 66, 72, 125–31 (Russia), and 54, 72–3 (Russian Civil War); C.J. Jenner, post to H–Minerva, 3 February 1999, on current research on women in the People's Army of Vietnam as special forces and sappers. For a discussion of earlier cases of women's involvement in combat, and an argument that women need to be aware of their share of the warrior tradition see David E. Jones, *Women Warriors: A History*, Brassey's, Washington, DC, 1997, passim.

6 John Erickson, 'Soviet women at war', in *World War 2 and the Soviet People: Selected Papers from the Fourth World Congress for Soviet and East European Studies, Harrogate, 1990*, John Garrard & Carol Garrard (eds), St Martin's Press, London, 1993, pp. 51, 53–8; Richard Overy, *Russia's War: Blood upon the Snow*, TV Books, New York, 1997, pp. 272–3. An earlier assessment, more critical than that of Erickson, is contained in Anne Eliot Griesse & Richard Stites, 'Russia: revolution and war' in *Female Soldiers—Combatants or Non-Combatants? Historical and Contemporary Perspectives*, Nancy Loring Goldman (ed.), Westport, 1982, pp. 68–78.

7 Erickson, 'Soviet women at war', p. 52.

8 Erickson, 'Soviet women at war', pp. 52, 59. On the use of these militias see Geoffrey Hosking, *A History of the Soviet Union 1917–1991*, final edn, Fontana Press, London, 1992, pp. 277–9.

9 Erickson, 'Soviet women at war', pp. 60–2; Shelley Saywell, *Women in War*, Viking, Markham, 1985, pp. 134, 137.

10 Erickson, 'Soviet women at war', p. 50.

11 Saywell, *Women in War*, pp. 131, 148.

12 Erickson, 'Soviet women at war', p. 63. Griesse & Stites, 'Russia: revolution and war', pp. 69, 72–3, argue that men had roles in some of these women's units but not all.

13 Erickson, 'Soviet women at war', pp. 63–4; see also the interviews in Saywell, *Women in War*, pp. 130–59. A collection of interviews with women veterans published in the USSR in 1989 has not yet been translated into English: Hodgson, 'The other veterans', *World War 2 and the Russian People*, pp. 91–2.

14 Erickson, 'Soviet women at war', pp. 64–5.

15 On prejudice and opposition see ibid, pp. 64, 67–9; Saywell, *Women in War*, p. 147; Griesse & Stites, 'Russia: revolution and war', pp. 72–3; on lack of provision for sanitary and medical facilities see Erickson, 'Soviet women at war', pp. 67, 69. The usual social policing rumours of alleged promiscuity at the front also followed female veterans: Hodgson, 'The other veterans', p. 94. Rogan claims pregnancy caused problems in Soviet partisan units but not in Yugoslav partisan units: Helen Rogan, *Mixed Company: Women in the Modern Army*, Putnam's, New York, 1981, pp. 85, 89.

16 Saywell, *Women in War*, pp. 183–4, suggests that objections by Jewish religious leaders played a part in the change in Israeli policy.

17 US General Accounting Office, National Security and International Affairs Division, *Report to the Honorable John W. Warner, U.S. Senate. Homosexuals in the Military: Policies and Practices of Foreign Countries*, Washington, June 1993, p. 39. On an analogous issue, Adam argues that 'only the Israeli military attracts the attention of American commentators' where the gays in the military question is concerned: Barry D. Adam, 'Anatomy of a panic: state voyeurism, gender politics, and the cult of Americanism', in *Gays and Lesbians in the Military: Issues, Concerns and Contrasts*, Wilbur J. Scott & Sandra Carson Stanley (eds), Aldine de Gruyter, New York, 1994, p. 113.

18 'Admiral backs women in combat', *The Age*, 22 February 1999, p. 6.

19 'Employment of women in the Australian Defence Force', DI (G) PERS 32–1, p. 1.

20 ibid, pp. 2–3. Women are also barred from work as surface finishers and electroplaters: New Zealand Defence Force policy is similar. Women are barred from service as Navy divers; there are no restrictions on women's service in the New Zealand Air Force; in the New Zealand Army women are excluded from combat positions such as the infantry, artillery, armour and Special Air Service: 24 March 1999, letter from Lieutenant Commander D.K. Washer, New Zealand Defence Force, p. 1.

21 'Frauen in der Bundeswehr', *http://www.bundeswehr.de/bundeswehr.frauen.vorwort.htm* (accessed 9 March 1999). Germany also has one of the more restrictive policies in place on male homosexuals: Paul Gade, David Segal & Edgar Johnson, 'The experience of foreign militaries', in *Out in Force: Sexual Orientation and the Military*, Gregory M. Herek, Jared B. Jobe & Ralph M. Carney (eds), University of Chicago Press, Chicago, IL, 1996, pp. 120–1.

22 There are some similarities between current French and Australian policies. Women in the French army may not be employed in areas with the possibility of direct and prolonged conflict with hostile forces: 'Le Dossier de la Semaine: la Féminisation dans les Armées', 31 October 1998, p. 40. In the Navy women are not employed in submarines, as carrier pilots and marine commandos: ibid, p. 43. Whittaker's 'We are not alone' claims that in the French Air Force women are not allowed in commando roles; French government material, however, states that all Air Force positions are open to women: ibid, p. 45.

23 UK policy is set out on *http://www.mod.uk/news/prs/056_99.htm* (accessed 9 March 1999); Whittaker, 'We are not alone'.

24 *http://www.dnd.ca/navy/marcom/tin/tin_e.htm#9* (accessed 9 March 1999); 'Vision 2010—the integrated Navy', 1998, *http://www.dnd.ca/navy/marcom/tin/summ_e.htm* (accessed 5 March 1999); 'Women might serve in new subs', 20 February 1999 report in the *Montreal Gazette*, H-Minerva post of 23 February 1999. The same report stated that women held 3.1 per cent of combat positions in the Canadian forces, up from 1 per cent in 1989.

25 In Belgium, women have access to all categories and units, and form 6.8 per cent of the armed forces: *http://www.mil.be/landmacht/paginas/gb/persone.htm* (accessed 5 March 1999). In Norway, women have been permitted to volunteer for service under the same conditions and access as men since 1979. The Norwegian Constitution contains provisions that would allow women to be conscripted: 'Norwegian defence', attachment to fax from Ms Monica O. Nagelgaard, Second Secretary, Royal Norwegian Embassy, Canberra, 9 March 1999, pp. 1, 4–5. Sweden, Finland and Denmark also have no restrictions on the positions which can be held by women in their armed forces: 'The defence of Sweden', *http://www.mil.se/intro/intro4_e.html* (accessed 9 March 1999), and telephone conversation with Colonel Nilsson, Swedish Defence Attaché, Canberra, 4 March 1999; 'Strength of the Defence Forces', *http://www.mil.f/english/s20.html* (accessed 9 March 1999); Kate Muir, *Arms and the Woman*, Coronet Books, London, 1993, p. 16. On Finnish policy see also 'Nr 194/1995 Lag om frivillig militärjänst för kvinnor', pp. 1–2, and *Bevärings-tjänsten*, Helsinki, 1998, pp. 52–3. The Bundeswehr homepage on women in NATO forces claims that restrictions exist in Belgium (submarines), Denmark (commandos and frogmen), and

the Netherlands (submarines and Marines): 'C. Frauen in den Streitkräften anderer Staaten, 3. Verwendungsmöglichkeiten für Frauen'. Whittaker, 'We are not alone', and the material provided by both the Belgian and Danish armed forces on their homepages indicates that there are no restrictions. Does this apparent discrepancy arise because, while all areas are open, women have not yet entered them all? In Norway and Denmark women are about 5 per cent of the armed forces: 'Women in the Canadian Defence Forces', 'The Press Room', *http://www.dnd.ca/ eng/archive/may98/women_b_e.htm* (accessed 5 March 1999). Since April 1998 Danish law allows the conscription of women, and Denmark's Ministry of Defence aims to increase the percentage of women in the armed forces to 20 per cent: 4 May 1999, letter from Ms Hanni M. Gredsted, Information and Cultural Officer, Royal Danish Embassy, Canberra.

26 On the greater opposition to these changes in the USA see Ilene Rose Feinman, 'Women warriors/women peacemakers: will the real feminists please stand up!', in *The Women and War Reader*, Lois Ann Lorentzen & Jennifer Turpin (eds), New York University Press, New York, (1998), pp. 135–7; Cynthia H. Enloe, 'The politics of constructing the American woman soldier', *Women Soldiers*, pp. 102–6. On the publicity given to US women's involvement in the Gulf War see Enloe, 'The politics of constructing the American woman soldier', pp. 81–110; Julie Wheelwright, '"It was exactly like the movies!" The media's use of the feminine during the Gulf War', *Women Soldiers*, pp. 111–34. For a generally positive assessment of this deployment see US General Accounting Office Report to the Secretary of Defense, July 1993, *Women in the Military: Deployment in the Persian Gulf War*, Washington, DC, 1993.

27 Quoted in Muir, *Arms and the Woman*, p. 140.

28 Canadian women constitute more than 10.8 per cent of the Canadian regular forces and 18.7 per cent of the reserve: 'Women in the Canadian Forces', February 1998, *http://www.dnd.ca/eng/archive/may98/women_b_e.htm* (accessed 5 March 1999). They have 2 per cent of noncommissioned members (NCMs) and 5.2 per cent of officers in combat arms from 1987 to 1996: Chief Land Staff, *Gender Integration Study: The Experience of Women Who Have Served in the Combat Arms*, *http://www.dnd.ca/eng/min/reports/cls-gis/sep-sum.htm* (accessed 5 March 1999). The Department of National Defence homepage contains the texts of various papers setting out Canadian policy: *http://www.dnd.ca/*

29 The Swedish armed forces, though, admit women only as officers: telephone conversation with Colonel Nilsson, Swedish Defence Attaché, Canberra, 4 March 1999.

30 Ireland (2.2 children per woman of child-bearing age) and Israel (2.6 children per Israeli Jewish woman of child-bearing age; the Israeli Arab birthrate is higher) are the two main exceptions: Uta Klein,

'War and gender: what do we learn from Israel?', *Women and War Reader*, p. 150. The Australian birthrate has been below replacement level (defined as 2.1 children per woman of child-bearing age) since 1975: Doug Cocks, *People Policy: Australia's Population Choices*, University of New South Wales Press, Sydney, 1996, p. 11. On the causes of the decline in Australia see Michelle Gunn, 'Boom or doom', *The Australian*, 2 April 1999, p. 15. On demographic trends in the Western world, see Massimo Livi-Bacci, *A Concise History of World Population*, Blackwell, Cambridge, MA, 1992, pp. 113, 115–17, 121–3. I am grateful to Professor David Garrioch for this reference.

31 On the fear of losses see Edward N. Luttwak, 'Towards post-heroic warfare', *Foreign Affairs*, May/June 1995, 74/3, p. 115; Edward N. Luttwak, 'A post-heroic military policy', *Foreign Affairs*, July/August 1996, 75/4, p. 39; Francis Fukuyama, 'Women and the evolution of world politics', *Foreign Affairs*, 77/5, September/October 1998, p. 38. This in turn reinforces the demilitarisation of the values of Western societies: Cathy Downes, *Social, Economic and Political Influences upon the Australian Army of the 1990s*, Strategic and Defence Studies Centre, Research School of Pacific Studies, Australian National University, Canberra, 1988, pp. 7–9.

32 Squeamishness about enemy civilian casualties, in Iraq and Yugoslavia for example, marks a departure from the readiness to accept this with area bombing in World War II.

33 For a general argument that the West is becoming a post-military society see Martin Shaw, *Post-Military Society: Militarism, Demilitarization and War at the End of the Twentieth Century*, Polity Press, Cambridge, 1991, passim. See also Downes, *Social, Economic and Political Influences upon the Australian Army of the 1990s*, pp. 7–9.

34 Shaw, *Post-Military Society*, pp. 86, 88, 100. Shaw argues that countries with a professional military are less likely to have had a crisis of legitimacy.

35 Male conscription in the Netherlands was suspended on 1 January 1997: 'Objectives and tasks of the Ministry of Defence', *htpp://www.mindef.nl/english/org1.htm* (accessed 9 March 1999).

36 National Organization of Women, 1990, National Board Policy: Women in the military, *http://www.now.org/issues/military/policies/wim.html* (accessed 5 March 1999); for a statement of the WEL position see Sarah Maddison, 'Right to fight is feminism's frontline', *The Weekend Australian*, 9–10 January 1999, p. 21.

37 On the right to access to combat positions as an aspect of full citizenship see Elisabetta Addis, Valeria E. Russo & Lorenza Sebesta, 'Introduction', *Women Soldiers*, p. xiv. For an argument highlighting the economic disadvantages to women of their exclusion see Elisabetta Addis, 'Women and the economic consequences of being a soldier', ibid, pp. 3–27.

38 On the campaign concerning rape as a war crime see Rhonda Copelon, 'Surfacing gender: reconceptualizing crimes against women in times of war', *Women and War Reader*, pp. 63–79. On the NOW campaigns on this issue and on the Taliban regime see 'Bring justice to victims of war-time rape in Bosnia', *http://www.now.org/issues/global/bosnia.html* (accessed 5 March 1999); Dawn Corbett, 'Women around the globe face threats to human rights', *http://www.now.org/nnt/fall–98/global.html* (accessed 5 March 1999); 'Stop the abuse of women and girls in Afghanistan!', *http://www.now.org/foundation/global/taliban.html* (accessed 5 March 1999); 'Hillary Clinton blasts Taliban', *The Age*, 6 March 1999, p. 20.

39 For example, Cynthia Enloe has argued that the US military was willing to contemplate an increased role for women in the military out of fear that an all-volunteer male army would be disproportionately composed of African–American men: Enloe, 'The politics of constructing the American woman soldier', pp. 85–6. Both at the time of the transfer to a volunteer army and since, concern has been expressed in the USA that such an army would be disproportionately composed of African–Americans. This has been attributed to a white fear of an army controlled by African–Americans: Martin Binkin & Mark J. Eitelberg with Alvin J. Schexnider & Marvin M. Smith, *Blacks and the Military*, Brookings Institute, Washington, DC, 1982, pp. 41–3, 81.

40 Gisela Kaplan, *Contemporary Western European Feminism*, UCL Press, London, 1992, pp. 153–6.

41 Among its slogans are 'feminism is no pacifism'. On its strength and militancy see ibid, pp. 249–52.

42 Addis et al., *Women Soldiers* 'Introduction', p. xxiii.

43 Background to the Italian policy is provided by Virginia Ilari, 'Penelope's web: female military service in Italy—debates and draft proposals, 1945–92', ibid, pp. 150–61.

44 Letter from Brigadier-General Valter Mauloni, Air Military and Naval Attaché, Embassy of Italy, Canberra, 12 March 1999.

45 The relevant article is Article 12 (a) subsection 4, which came into force in June 1968, and states, after indicating that women may be called on to meet civil defence requirements, 'they may on no account render service involving the use of arms': 'Appendix Translation of the Basic Law of the Federal Republic of Germany', in *The Constitution of the Federal Republic of Germany: Essays on the Basic Rights and Principles of the Basic Law with a Translation of the Basic Law*, Ulrich Karpen (ed.), Nomos Verlagsgesellschaft, Baden-Baden, 1988, p. 232.

46 Jeff M. Tuten, 'Germany and the world wars', *Female Soldiers— Combatants or Non-Combatants?*, pp. 47, 52–7.

47 'Frauen in der Bundeswehr 2. Die Diskussion seit Ende der siebziger Jahre', *http://www.bundeswehr.de/bundeswehr/frauen/A-Disk70.htm* (accessed 9 March 1999). In 1997 women were only 0.9 per cent of German soldiers:

Heidemarie Kemnitz, 'Mädchen und Militär: "Weiblichkeit" im Diskurs um nationale Frauendienste, Wehrerziehung für Mädchen und Frauen in der Bundeswehr', *Feministische Studien*, 16/1, 1998, p. 70.

48 Whittaker, 'We are not alone'.

49 The reasons for the constitutional ban are not clear. Was it an attempt to protect German women from war, a symptom of general anti-war feeling or a case of the 'remasculinisation' of Germany after the Occupation? As the 'Basic Law' has often been changed, the existence of the ban is not an immovable obstacle. For German women's attitudes: 'Frauen ans Gewehr? Das Brigitte-Gespräch', *Brigitte*, 14/79, pp. 81–5; Kemnitz, 'Mädchen und Militär', pp. 69–85. On the weakness of German feminism: Kaplan, *Contemporary Western European Feminism*, pp. 103, 115.

50 Kaplan, *Contemporary Western European Feminism*, p. 117.

51 There are defensive as well as sexist tones in the Bundeswehr's account of the policy of other NATO countries: 'C. Frauen in den Streitkräften anderer Staaten'.

52 Richard A. Preston, Alex Roland & Sydney F. Wise, *Men in Arms: A History of Warfare and its Interrelationships with Western Society*, 5th edn, Harcourt Brace Jovanovich College, Fort Worth, TX, 1990, p. 331.

53 For a discussion of these issues see Patricia Pearson, *When She Was Bad: How Women Get Away with Murder*, Virago, London, 1998, passim. Pearson argues that social (and feminist) stereotypes about women's alleged innate non-violence blind us to the clear evidence of their use of violence. In addition, recent research has suggested that the role testosterone plays in aggressive behaviour has been exaggerated: ibid, pp. 8–9; Barbara Ehrenreich, 'The real truth about the female', *Time*, 8 March 1999, p. 38.

54 Pearson, *When She Was Bad*, pp. 31–2. Pearson's examples are mainly taken from North America. For similar developments in Britain see Lisa Brinkworth, 'Twisted sisters', *Harpers & Queen*, April 1997, pp. 131–3; and in Australia: Paulyne Pogonelska, 'Females in a fury', *The Age*, 4 February 1999, p. 16.

55 Pearson, *When She Was Bad*, p. 16; Philip Jacobson, 'Blood sisters', *The Australian* Magazine, 2–3 March 1996, pp. 12–18.

56 On this issue in the Australian context see Downes, *Social, Economic and Political Influences upon the Australian Army of the 1990s*, pp. 7–10, 16–17.

57 Downes argues that women are still a largely untapped resource in the Australian military: ibid, p. 8.

58 On the 'gender gap' on attitudes to defence and foreign policy: Robin Morgan, 'Complicated trust', *Ms Magazine*, January/February 1991, p. 1; Fukuyama, 'Women and the evolution of world politics', pp. 24–40 and the various 'responses' to his article, *Foreign Affairs*, 78/1, January/February 1999, pp. 118–29. Shaw argues that the

'gender gap' in Britain is not as great as in the USA: Shaw, *Post-Military Society*, pp. 150, 211–12.

59 See as a comparison the arguments of A. Philip Randolph of the Pullman Porter's Union in 1948 on the segregated US military: quoted in Garry L. Rolson & Thomas K. Nakayama, 'Defensive discourse: blacks and gays in the U.S. military', *Gays and Lesbians in the Military*, p. 124. On the economic costs to women of exclusion from, or restriction of opportunities in, the military see Addis et al., 'Introduction', pp. xiv, xx; Elisabetta Addis, 'Women and the economic consequences of being a soldier', ibid, pp. 3–27.

60 Hodgson, 'The other veterans', pp. 78–9.

61 See for example the accounts in Saywell, *Women in War*, pp. 152–3, of women carrying men more than double their weight in battle conditions and pp. 155–6, 158 of women in close combat.

62 Quoted in Hodgson, 'The other veterans', p. 82.

63 Chief Land Staff, *Gender Integration Study*, http://www.dnd.ca/eng/min/ reports/cls-gis/dav–4.htm para. 14 (accessed 5 March 1999).

64 Chief Land Staff, *Gender Integration Study*, ibid, para. 18, and http://www.dnd.ca/eng/min/reports/cls-gis/sep–3.htm para. 19 (accessed 5 March 1999).

65 Ehrenreich, 'The real truth about the female', p. 38; Burton, *Women in the Australian Defence Force*, ES–25. On physical differences between the sexes see also Judith Lorber, *Paradoxes of Gender*, Yale University Press, New Haven, CT, 1994, pp. 41–4, 49–51. On the whole issue of upper-body strength see also Judith Hicks Stiehm, *Bring Me Men and Women: Mandated Change at the U.S. Air Force Academy*, University of California Press, Berkeley, CA, 1981, ch. 6, passim.

66 'Diagnosis female', *Time*, 8 March 1999, p. 46; email by Michael Yared to Eleanor Hancock, 2 February 1999, containing the text of a press report by Rowan Scarborough, 'Study: Annapolis women's knees more injury prone', *The Washington Times*, 12 July 1998.

67 See the recommendations made by the participants in the Canadian study: 'Annex F interview participant recommendations', Chief Land Staff, *Gender Integration Study*, http://www.dnd.ca/eng/min/reports/cls-gis/ sep–6.htm and http://www.dnd.ca/eng/min/reports/cls-gis/dav-sum.htm (accessed 5 March 1999).

68 Ehrenreich, 'The real truth about the female', p. 38. For the opposite view see the works of Brian Mitchell, such as *Weak Link: The Feminization of the American Military*, Regnery Gateway, Washington, DC, 1989, passim; and *Women in the Military: Flirting with Disaster*, Regnery, Washington, DC, 1998, passim. Mitchell's work is open to several criticisms. He fails to distinguish in his criticisms of the gender-integrated military between lasting difficulties and problems of transition; he blames women in the forces for the mistaken decisions of male military decision makers during the beginnings of

integration; and his homophobia leads him to disparage the group of
women in the military who, he considers, make the best women
soldiers—lesbians: Mitchell, *Weak Link*, pp. 178–82. For a critique
of Mitchell see Paul E. Roush, 'Combat exclusion: military necessity
or another name for bigotry', *MINERVA: Quarterly Report on Women
and the Military*, VIII/3, 1990, pp. 6–9, 12–13.

69 Chief Land Staff, *Gender Integration Study*, http://www.dnd.ca/eng/min/
reports/cls-gis/dav–4.htm para. 30, and http://www.dnd.ca/eng/min/reports/
cls-gis/sep–3.htm para. 21–22, 27 (accessed 5 March 1999); Muir, *Arms
and the Woman*, pp. 99–111.

70 Rogan, *Mixed Company*, p. 68. Was this the reason for women's
better results in tests for the Mercury astronauts? The Mercury
program is an example of a failure, due to sexist prejudice, to draw
on women's physiological strengths in withstanding the stresses of
space flight: Bob Graham, 'Right stuff wrong sex', *The Australian
Magazine*, 23–24 January 1999, pp. 11–17. D'Ann Campbell notes
that reluctance to devise tests of physical fitness where women might
excel continues at West Point: D'Ann Campbell, 'The spirit run and
football cordon: a case study of female cadets at the U.S. Military
Academy', in *Forging the Sword: Selecting, Educating, and Training
Cadets and Junior Officers in the Modern World*, Elliott V. Converse III
(ed.), Military History Symposium Series of the United States Air
Force Academy, vol. 5, Imprint Publications, Chicago, IL, 1998,
p. 239

71 Hodgson, 'The other veterans', p. 92.

72 Felicja Karay, 'Women in the forced-labor camps', in *Women in the
Holocaust*, Dalia Ofer & Lenore J. Weitzman (eds), Yale University
Press, New Haven, CT, 1998, pp. 306–7.

73 D'Ann Campbell, 'The spirit run and football cordon', p. 241; Lynne
Hanley, 'To kill and be killed' (review of Carol Barkalow & Andrea
Raab, *In the Men's House*), *The Women's Review of Books*, 8/6, March
1991, p. 3. I am grateful to Mr Tony Wood for providing this reference.

74 Addis et al., *Women Soldiers*, 'Introduction', pp. xix–xx.

75 Enloe, 'The politics of constructing the American woman soldier',
pp. 99–100; Binkin, Eitelberg et al., *Blacks and the Military*, p. 37.

76 Muir, *Arms and the Woman*, pp. 123–4, 126, notes that military
psychology has been designed to motivate men and it will need to
change; see also the comments on training women of former WACs
in Rogan, *Mixed Company*, pp. 177–9. Current US training techniques may be particularly ill-suited to a gender-integrated military
because they play on men's fears of femininity and homosexuality:
Adam, 'Anatomy of a panic', pp. 111–12; Gwynne Dyer, *War*, Crown
Publishers, New York, 1985, pp. 111, 123–5.

77 Guenter Lewy, 'The American experience in Vietnam', *in Combat
Effectiveness: Cohesion, Stress and the Volunteer Military*, Sam C. Sarkesian

(ed.), Sage, Beverley Hills, CA, 1980, pp. 95–6; Michael R. Kauth & Dan Landis, 'Applying lessons learned from minority integration in the military', *Out in Force*, p. 89. On research into bonding more generally see Sarkesian, *Combat Effectiveness*, passim; and the survey of literature in Robert J. MacCoun, 'Sexual orientation and military cohesion: a critical review of the evidence', *Out in Force*, pp. 157–64, 172.

78 'The training and employment environment: male combat arms culture', Chief Land Staff, *Gender Integration Study*, http://www.dnd.ca/eng/min/reports/cls-gis/dav–4.htm para. 14 (accessed 5 March 1999). Author's emphasis.

79 Canadian authorities have termed these actions 'gender harassment': Chief Land Staff, *Gender Integration Study*, http://www.dnd.ca/eng/min/reports/cls-gis/dav–4.htm paras 19 and 30 (accessed 5 March 1999).

80 'Interview results: beyond physical ability', ibid, para. 14; http://www.dnd.ca/eng/min/reports/cls-gis/sep–4.htm, para. 30 (accessed 5 March 1999).

81 Stiehm, *Bring Me Men and Women*, passim. These prejudices are also at work in some of the frequent public objections to the use of women in combat and combat-related roles, which use a variation of the 'no true Scotsman' fallacy—women cannot meet the standards, and if they do they are not real women: Anthony Flew, *Thinking about Thinking (Or, Do I Sincerely Want to be Right?)*, Fontana, London, 1985, pp. 47–54; L. Susan Stebbing, *Thinking to Some Purpose*, Penguin, Harmondsworth, 1959, pp. 171–2.

82 Terry quotes Major-General Jeanne Holm: 'to be overly concerned about the welfare and safety of military women while at the same time accepting as inevitable the obliteration of literally millions of civilian women and children in the event of a nuclear confrontation is ludicrous': Ray Terry, *Women in Khaki: The Story of the British Woman Soldier*, Columbus, London, 1988, p. 245.

83 'Lieutenant Vivian Bullwinkel', in *The Oxford Companion to Australian Military History*, Peter Dennis, Jeffrey Grey, Ewan Morris, Robin Prior with John Connor (eds), Oxford, Melbourne, 1995, p. 128.

84 It has been claimed that on 24 April 1945, that is, at 'five minutes before midnight', Hitler did order the drafting of some women in central Berlin into Mohnke's battle group: Anthony Read & David Fisher, *The Fall of Berlin*, Pimlico, London, 1993, p. 387, but the authors give no source for their claim.

85 Dennis L. Bark & David R. Gress, *A History of West Germany, Vol. 1: From Shadow to Substance, 1945–1963*, 2nd edn, Blackwell, Oxford, 1993, pp. 1, 32–7, 40; Douglas Botting, *In the Ruins of the Reich*, Grafton Books, London, 1985, p. 228; Overy, *Russia's War*, pp. 308, 310; Christopher Duffy, *Red Storm on the Reich: The Soviet March on Germany*, 1945, Atheneum, New York, 1991, pp. 275–6.

86 Bark & Gress, *A History of West Germany, Vol. 1*, pp. 43–5.

87 Indeed, British investigations of the legal position of the women's services found that, aside from medical staff, there is no such thing in international law as noncombatant status in the armed forces: Terry, *Women in Khaki*, pp. 212–13, 218.

88 Gunther E. Rothenberg, 'The age of Napoleon', in *The Laws of War: Constraints on Warfare in the Western World*, Michael Howard, George J. Andreopoulos & Mark R. Shulman (eds), Yale University Press, New Haven, CT, 1994, p. 93.

89 On the lack of training in combat of noncombat personnel in Northern Ireland and the risks they ran see Sarah Ford, *One Up: A Woman in Action in the SAS*, Harper Collins, London, 1997, pp. 171–5. On the desire of women in the WRAC to have the means and training to protect themselves: Terry, *Women in Khaki*, pp. 213–16.

90 Stiehm, *Bring Me Men and Women*, p. 158.

91 Andy McNab, *Bravo Two Zero*, Corgi Books, London, 1997, pp. 16, 200, 279, 323.

92 Stiehm, *Bring Me Men and Women*, pp. 296–7.

93 Muir, *Arms and the Woman*, p. 69. In the Canadian Navy, on the other hand, women are quartered separately: 'Vision 2010—the integrated Navy', *http://www.dnd.ca/navy/marcom/tin/tin_e.htm#10* paras 68–9 (accessed 5 March 1999).

94 It should be noted, however, that since 1987 the Canadian military have required all women to agree to combat when they sign up without women's applications falling off. Applications for the Royal Navy from women went up 30 per cent once they were allowed to go to sea: Muir, *Arms and the Woman*, p. 114.

95 Burton, *Women in the Australian Defence Force*, ES–10. The evidence of behaviour of men towards women under extreme conditions suggests that protective instincts are socially constructed: Michael Unger, 'Women in the Lodz Ghetto', *Women in the Holocaust*, pp. 136–9.

96 Kathy Dobie, 'A few nice men', *US Vogue*, February 1993, p. 299.

97 Telephone conversation with Colonel Nilsson, Swedish Defence Attaché, Canberra, 4 March 1999. The US Navy and Marines have been more conservative than the USAAF: Enloe, 'The politics of constructing the American woman soldier', pp. 92–4.

98 On sexual harassment in the Australian context: *Government Response to 'Facing the Future Together: Report on Sexual Harassment in the Australian Defence Force by the Senate Standing Committee on Foreign Affairs, Defence and Trade'*, Australian Government Publishing Service, Canberra, December 1994.

99 Burton, *Women in the Australian Defence Force*, ES–8, ES–29 – ES–30.

100 Ultimately training must be gender-integrated, even if as an interim measure some aspects might be separate. On the US debates about integrated training see the Kassebaum Baker report on genderintegrated

training: *http://www.defenselink.mil/pubs/g.t* (accessed 9 March 1999); Rowan Scarborough, 'Coed military training saluted', *The Washington Times*, 16 March 1999, posted to H-Minerva, 16 March 1999.

101 Smith & McAllister, 'The changing military profession', pp. 387–8. In this context one might also note the greater acceptance of women in the combat arms of the Canadian Reserve Force: Chief Land Staff, *Gender Integration Study*, *http://www.dnd.ca/eng/min/reports/cls-gis/may–4.htm* paras 21–3 (accessed 5 March 1999).

102 Similar reasons may have made it easier for the Netherlands and Belgian military as they made the transition from a conscript to a volunteer military.

103 On the mistakes made, and on opposition to the policies see Rogan, *Mixed Company*, pp. 160, 171–3, 177, 215; Stiehm, *Bring Me Men and Women*, passim, but especially, pp. 111–13, 121–2, 128 (initial ban on women over 6 feet tall!). D'Ann Campbell has recently argued that conditions for women at West Point have improved and has noted the impressive programs now being developed by the US military against all forms of discrimination: D'Ann Campbell, 'The spirit run and football cordon', pp. 237–47.

104 Muir, *Arms and the Woman*, pp. 123–4, 131.

105 Erickson, 'Soviet women at war', pp. 60–1.

106 Eleanor Hancock, *The National Socialist Leadership and Total War, 1941–5*, St Martin's Press, New York, 1991, pp. 94, 135, 142–3, 178, 183, note 14 p. 294, note 32 p. 297.

107 For an example of the psychology of the pioneers see Sarah Ford, *One Up*, passim; Chief Land Staff, *Gender Integration Study*, *http://www.dnd.ca/eng/min/reports/cls-gis/dav–4.htm* para. 35 (accessed 5 March 1999).

108 Judith Hicks Stiehm has argued that US policy on homosexuals in the military has unfairly affected all women in the armed forces: Judith Hicks Stiehm, 'The military ban on women and the Cyclops effect', in *Gays and Lesbians in the Military: Issues, Concerns and Contrasts*, Wilbur J. Scott & Sandra Carson Stanley (eds), Aldine de Gruyter, New York, 1994, pp. 149–62.

109 General Colin Powell, speech of 10 November 1998, *http://www.mod.uk/policy/equalops/poweladd.htm* (accessed 5 March 1999).

110 Stiehm, *Bring Me Men and Women*, pp. 237–8; Muir, *Arms and the Woman*, p. 175; Rogan, *Mixed Company*, p. 212; Chief Land Staff, *Gender Integration Study*, *http://www.dnd.ca/eng/min/reports/cls-gis/dav–4.htm* para. 15, and *http://www.dnd.ca/eng/min/reports/cls-gis/sep–3.htm* para. 19 (accessed 5 March 1999). For its failure to be enforced in Australia see Burton, *Women in the Australian Defence Forces*, ES3–5.

111 Powell, *http://www.mod.uk/policy/equalops/poweladd.htm* (accessed 5 March 1999). Despite General Powell's own successful career, it has been argued that there is a 'glass ceiling' for almost all African–American officers:

Kauth & Landis, 'Applying lessons learned from minority integration in the military', *Out in Force*, p. 91. On opposition to integration in the 1940s: ibid, pp. 93–4. Their overall view of the measures is less upbeat than General Powell's: ibid, pp. 88–91.

112 Stiehm quotes Charlotte Perkins Gilman's comment: 'The thing a woman is most afraid to meet on a dark street is her natural protector': Stiehm, *Bring Me Men and Women*, p. 299.

113 Jill Radford & Diana E.H. Russell (eds), *Femicide: The Politics of Woman Killing*, Twayne Publishers, New York, 1992.

114 Muir, *Arms and the Woman*, p. 130; Ehrenreich, 'The real truth about the female', p. 42.

115 Saywell, *Women in War*, pp. 148–9, 150–1. As European urban terrorists, in fact, women showed greater commitment and fought harder: Mark Thornton, University of Western Sydney, 'The right to be bad: Italian terrorism and the politics of gender', paper given to the Australian Historical Association Conference 1992, p. 15.

116 Hodgson, 'The other veterans', p. 93.

117 'Today's Europe: fewer threats but less peace', *http://www.mil.se/intro/intro2_e.html* (accessed 9 March 1999).

118 For a discussion of some of the requirements of this see Luttwak, 'Towards post-heroic warfare', pp. 109–22; Luttwak, 'A post-heroic military policy', pp. 33–44. For an argument that insufficient use has been made of women in peacekeeping, due in part to sexism within the United Nations, see Janet Beilstein, 'The expanded role of women in United Nations peacekeeping', *Women and War Reader*, pp. 140–7.

119 Chief Land Staff, *Gender Integration Study: An Exploration of Reserve Member Perceptions of Regular Force Combat Arms Employment*, *http://www.dnd.ca/eng/min/reports/cls-gis/may–4.htm* paras 21 and 22 (accessed 5 March 1999). Cheeseman & Hall also raise the question whether current military training in the ADF is preparing any officers for the military of the future: Graeme Cheeseman & Robert A. Hall, *Preparing for Australia's Military After Next: The Price Report and a 'New Model' Australian Defence Force Academy*, Australian Defence Studies Centre, Canberra, 1997, pp. 13–25, 26–41.

120 See for examples: Stiehm, *Bring Me Men and Women*, pp. 199, 264.

121 Azar Nafisi, 'The veiled threat in Iran', *The Australian*, 6–7 March 1999, pp. 32–3.

122 Hugh Smith, 'Women in the military: developments, debates and dilemmas', unpublished paper, 1998, p. 12.

123 Stiehm, *Bring Me Men and Women*, pp. 298–9.

124 'You must tell your children, Putting modesty aside, That without us, without women, There would've been no spring in 1945' as the poem by the Soviet anti-aircraft gunner Nouna Alexandra Smirnova puts it: quoted in Janet Howarth, 'Women at war', *The Oxford*

Companion to the Second World War, I.C.B. Dear & M.R.D. Foot (eds), Oxford University Press, Oxford, 1995, p. 1282.
125 ibid, p. 294.
126 Binkin, Eitelberg et al., *Blacks in the Military*, pp. 116–19.
127 Stiehm, *Bring Me Men and Women*, pp. 293–6.
128 Nancy F. Cott, 'Marriage and women's citizenship in the United States, 1830–1934', *The American Historical Review*, 103/5, December 1998, pp. 1440–74.
129 Military service is seen as an integral and normative part of citizenship: Donald H. Horner, Jr & Michael T. Anderson, 'Integration of homosexuals into the armed forces: racial and gender integration as a point of departure', *Gays and Lesbians in the Military*, p. 250; Stiehm, *Bring Me Men and Women*, pp. 297–9; Shaw, *Post-Military Society*, pp. 174–7; Binkin, Eitelberg et al., *Blacks in the Military*, pp. 25–6.
130 Cott, 'Marriage and women's citizenship in the United States, 1830–1934', p. 1473.

CHAPTER 11

1 S.J. Deitchman, *Quantifying the Military Value of Training for System and Force Acquisition Decisions: An Appreciation of the State of the Art*, IDA Paper P–2881, Institute for Defense Analysis, Alexandria, VA, 1993.
2 P.F. Gorman, *The Military Value of Training*, IDA Paper P–2515, Institute for Defense Analysis, Alexandria, VA, 1990.
3 ibid.
4 ibid, p. 15.
5 B.W. Hallmark & J.C. Crowley, *Company Performance at the National Training Center*, Rand, Washington, DC, 1997.
6 ibid, p. 48.
7 ibid, p. xvi.
8 Deitchman, *Quantifying the Military Value of Training*.
9 T.J. Thompson, R.J. Pleban & P.J. Valentine, *Determinants of Effective Unit Performance: Battle Staff Training and Synchronization*, US Army Research Institute for the Behavioral and Social Sciences, Alexandria, VA, 1994.
10 A.T. Welford, *Fundamentals of Skill*, Methuen, London, 1967.
11 J.M. Flach, 'Situation awareness: proceed with caution', *Human Factors*, vol. 37, no. 1, 1995, pp. 149–57.
12 See M.R. Endsley, 'Towards a theory of situation awareness in dynamic systems', *Human Factors*, vol. 37, no. 1, 1995, pp. 32–64; M.R. Endsley, 'The role of situation awareness in naturalistic decision making', in *Naturalistic Decision Making*, C.E. Zsambok & G. Klein (eds), Lawrence Earlbaum, Mahwah, NJ, 1997.

13 A.T. Welford, 'Mental work-load as a function of demand, capacity, strategy and skill', *Ergonomics*, vol. 21, no. 3, 1978, pp. 151–67.
14 R.E. Geiselman & M.G. Samet, 'Summarizing military information: an application of schema theory', *Human Factors*, vol. 22, no. 6, 1980, pp. 693–705.
15 G. Klein, 'The recognition-primed decision (RPD) model: looking back, looking forward', C.E. Zsambok & G. Klein (eds), *Naturalistic Decision Making*, Lawrence Earlbaum, Mahwah, NJ, 1997.
16 M.S. Cohen, J.T. Freeman & B.B. Thompson, 'Training the naturalistic decision maker', in Zsambok & Klein (eds), *Naturalistic Decision Making*.
17 Geiselman & Samet, 'Summarizing military information: an application of schema theory', pp. 693–705.
18 Hallmark & Crowley, *Company Performance at the National Training Center*.
19 L. McAulay, *The Battle of Coral,* Arrow, Sydney, 1988, p. 315.
20 H. Simpson, *Evaluating Large-Scale Training Simulations*, Defense Manpower Data Center, Washington, DC, 1998.
21 J.R. Bondonella, M.W. Lewis, L.P. Steinberg, G. Shin-K Park, D.G. Levy, E. Ettedgui, D.M. Oaks, J.M. Sollinger, J.D. Winkler, J.M. Halliday & S. Way-Smith, *Microworld Simulations for Command and Control Training of Theatre Logistics and Support Staffs: A Curriculum Strategy*, Rand, Washington, DC, 1999.
22 D. Rowland, 'The effect of combat degradation on the urban battle', *Journal of the Operational Research Society*, vol. 42, no. 7, 1991, pp. 543–53.
23 S. Horowitz, J. Orlansky, T.C. Tillson, T.C. Gemeles, H.J. Gilman, C. Hammon & H.M. Hoyler, *Unit Training in the Gulf War*, IDA Paper P–3087, Institute for Defense Analysis, Alexandria, VA, 1995.
24 Land Commander Australia, *Training Directive,* Sydney, 1998, (restricted).
25 T. Lewman, 'A conceptual framework for measuring unit performance', *Determinants of Effective Unit Performance*, R.F. Holz, J.H. Hiller, J.H. & H.H. McFann (eds), US Army Research Institute, Alexandria, VA, 1994.
26 M.R. McClusky, J.E. Fowlkes, L.G. Pierce & D.J. Dwyer, 'Measurement of command/control staff performance in tactical training environments', Proceedings of the Interservice/Industry Training, Simulation and Education Conference, Orlando, FA, 1998.
27 Australian Army, *Land Warfare Doctrine 1: The Fundamentals of Land Warfare*, Doctrine Wing, CATDC, Puckapunyal, 1998, pp. 2-2.
28 Hallmark & Crowley, *Company Performance at the National Training Center*.
29 J.A. Shaw, 'Psychodynamic considerations in the adaption to combat', in *Contemporary Studies in Combat Psychiatry*, G. Belenkey (ed.), Greenwood Press, London, 1987.

30 J.W. Appel & G.W. Beebe, 'Preventive psychiatry', *Journal of the American Medical Association*, vol. 131, no. 18, 1946, pp. 1469–76.

31 D.R. Haslam, M.F. Allnutt, D.E. Worsley, D. Dunn, J. Abraham, J. Few, S. Labuc & D.J. Lawrence, *The Effect of Continuous Operations upon the Military Performance of the Infantryman (Exercise Early Call)*, Army Personnel Research Establishment, Report 2/77, Farnborough, 1977.

32 G.R. Frank, 'Continuous operations', *ARMOR*, March/April 1982, pp. 19–23; G. Belenky, T. Balkin, G. Krueger, D. Headley & R. Solick, *Effects of Continuous Operations (CONOPS) on Soldier and Unit Performance, Phase 1: Review of the Literature*, US Army Combined Arms Combat Development Activity, Fort Leavenworth, 1986; G. Belenky, T. Balkin, G. Krueger, D. Headley & R. Solick, *Effects of Continuous Operations (CONOPS) on Soldier and Unit Performance: Review of the Literature and Strategies for Sustaining the Soldier in CONOPS*, Walter Reed Army Institute of Research, Bethesda, MD, 1987.

33 M.F. Allnutt, D.R. Haslam, M.H. Rejman & S. Green, *Sustained Performance and Some Effects on the Design and Operation of Complex Systems*, Army Personnel Research Establishment, Ministry of Defence, London, 1990.

34 C. Idzikowski, A.D. Baddeley, 'Fear and dangerous environments', in *Stress and Fatigue in Human Performance*, G.R. Hockey (ed.), Wiley, New York, 1983, pp. 123–44.

35 M.M. Berkun, 'Performance decrement under psychological stress', *Human Factors*, vol. 6, 1964, pp. 21–30; M.M. Berkun, H.M. Bialek, R.P. Kern & K. Yagi, 'Experimental studies of psychological stress in man', *Psychological Monographs*, vol. 9, no. 534, 1962.

36 D.C. Glass & J.E. Singer, *Urban Stress*, Academic Press, New York, 1972; Y. Gal-or, G. Tenenbaum, D. Furst & M. Shertzer, 'Effect of self-control and anxiety on training performance in young and novice parachuters', *Perceptual and Motor Skills*, vol. 60, 1985, pp. 743–6.

37 J. Griffith, 'The army's new unit personnel replacement and its relationship to unit cohesion and social support', *Military Psychology*, vol. 1, 1989, pp. 17–34; M. Steiner & M. Neumann, 'Traumatic neurosis and social support in the Yom Kippur War returnees', *Military Medicine*, vol. 143, 1978, pp. 866–8; A. Tziner & Y. Vardi, 'Effect of command style and group cohesiveness on the performance effectiveness of self-selected tank crews', *Journal of Applied Psychology*, vol. 67, 1982, pp. 769–75.

38 B.G. Kanki, 'Stress and air-crew performance: a team-level perspective', in *Stress and Human Performance*, J.E. Driskell & E. Salas (eds), Lawrence Erlbaum, Mahwah, NJ, 1996, pp. 127–62.

39 Deitchman, *Quantifying the Military Value of Training*.

40 A. Kellett, 'The soldier in battle: motivational and behavioral aspects of the combat experience', in *Psychological Dimensions of War*, B. Glad (ed.), Sage Publications, Newbury Park, CA, 1990, p. 220.

41 S.A. Stouffer, *The American Soldier: Combat and Its Aftermath, Studies in Social Psychology in World War II*, Vol. 2, Princeton University Press, Princeton, NJ, 1949.

42 Kanki, 'Stress and air-crew performance: a team-level perspective', pp. 127–62.

43 F.H. Norris & S.A. Murrell, 'Prior experience as a moderator of disaster impact on anxiety symptoms in older adults', *American Journal of Community Psychology*, vol. 16, 1998, pp. 665–83.

44 R.H. Ahrenfeldt, *Psychiatry in the British Army in the Second World War*, Routledge & Kegan Paul, London, 1958.

45 D. Meichenbaum, *Stress Inoculation Training*, Pergamon, New York, 1985.

46 C.A. Bowers, J.L. Weaver & B.B. Morgan, 'Moderating the performance effects of stressors', in *Stress and Human Performance*, J.E. Driskell & E. Salas (eds), Lawrence Erlbaum, Mahwah, NJ, 1996, pp. 163–92.

47 Kanki, 'Stress and air-crew performance: a team-level perspective', pp. 127–62.

48 J.M. Orasnu & P. Backer, 'Stress and military performance', in J.E. Driskell & E. Salas (eds), *Stress and Human Performance*, pp. 89–125.

49 R.A. Wertkin, 'Stress-inoculation training: principles and applications', *Journal of Contemporary Social Work*, vol. 66, 1985, pp. 611–16.

50 T. Saunders, J.E. Driskell, J.H. Johnston & E. Salas, 'The effect of stress inoculation training on anxiety and performance', *Journal of Occupational Health Psychology*, vol. 1, 1996, pp. 170–86.

51 ibid, p. 180.

52 Orasnu & Backer, 'Stress and military performance', pp. 89–125

53 R.B. Zajonc, 'Social facilitation', *Science*, vol. 149, 1965, pp. 269–74.

54 J.E. Driskell & E. Salas, 'Overcoming the effects of military training on military performance: human factors, training and selection strategies', in *Handbook of Military Psychology*, R. Gal & A.D. Mangelsdorff (eds), John Wiley, New York, 1991.

55 J.E. Driskell, R.P. Willis & C. Cooper, 'Effect of overlearning on retention', *Journal of Applied Psychology*, vol. 77, 1992, pp. 615–22.

56 R.L. Helmreich & H.C. Foushee, 'Why crew resource management? The history and status of human factors training progress in aviation', in *Crew Resource Management*, E.L. Weiner, B.G. Kanki and R.L. Helmreich (eds), Academic Press, New York, 1993, pp. 3–45.

57 N. Friedland & G. Keinan, 'Training effective performance in stressful situations: three approaches and implications for combat training', *Military Psychology*, vol. 4, 1992, p. 159.

58 ibid.

59 ibid, p. 172.

60 S. Milgram, 'Behavioral study of obedience', *Journal of Abnormal and Social Psychology*, vol. 67, 1963, pp. 371–8; C. Haney, W.C. Banks &

P.G. Zimbardo, 'Interpersonal dynamics in a simulated prison', *International Journal of Criminology and Penology*, vol. 1, 1973, pp. 69–97.

61 G. Keinan & N. Friedland, 'Training effective performance under stress: queries, dilemmas and possible solutions', in Driskell & Salas (eds), *Stress and Human Performance*, pp. 257–77.

62 A.C. McFarlane & R. Yehuda, 'Resilience, vulnerability and the course of post-traumatic reactions', in *Traumatic Stress: The Effects of Overwhelming Experience on Mind, Body and Society*, B.A. van der Kolk, A.C. McFarlane & L. Weisaeth (eds), The Guilford Press, New York, pp. 155–81.

63 D. Grossman, *On Killing*, Little Brown, London, 1996.

64 F. Skowronski, Overseas CTC visit report, Australian Army, 1998.

65 Hallmark & Crowley, *Company Performance at the National Training Center*.

66 T.W. Lucas, S.C. Bankes & P. Vye, *Improving the Analytic Contribution of Advanced Warfighting Experiment*, Documented Briefing DB–207-A, Rand, Washington, DC, 1998.

67 D. Rooney, V. Kallmeier & G. Stevens, *Mission Command and Battlefield Digitization: Human Sciences Considerations*, Report no. DERA/CHS/HS3/CR980097/1.0, DERA, Centre for Human Sciences, Farnborough, 1998.

68 D. Rowland, 'The effect of combat degradation on the urban battle', *Journal of the Operational Research Society*, vol. 42, no. 7, 1991, pp. 543–53.

69 R.W. Pew & A.S. Mavor (eds), *Representing Human Behavior in Military Simulations: Interim Report*, National Academy Press, Washington, DC, 1997.

70 Rooney, Kallmeier & Stevens, *Mission Command and Battlefield Digitization*.

CONCLUSION

1 Gerald F. Linderman, *Embattled Courage: The Experience of Combat in the American Civil War*, The Free Press, New York, 1987, p. 1.

2 Pyle, *Here Is Your War*, pp. 152–3.

3 Cited in Denis Winter, *Death's Men: Soldiers of the Great War*, Allen Lane, London, 1978, p. 267, fn 2.

4 Du Picq, *Battle Studies*, p. 128.

5 Joanna Bourke, *An Intimate History of Killing: Face-to-Face Killing in the Twentieth Century*, Granta Books, London, 1999, p. 1.

6 Tim O'Brien, *The Things They Carried*, Harper Collins, New York, 1990, p. 76.

7 John Keegan, *A History of Warfare*, Hutchinson, London, 1993, p. 76.

8 ibid.

9 Charles Moskos, *Air Force Times*, 2 November 1992.
10 Statement by Brigadier General Thomas V. Draude, in Association of the United States Army, *Women in Combat. Report to the President: The Presidential Commission on the Assignment of Women in the Armed Forces*, Brassey's, Washington, DC, 1993, p. 105.
11 S.L.A. Marshall, 'Salesmanship for the Army', 20 May 1957, in Roger J. Spiller (ed.), *S.L.A. Marshall at Leavenworth: Five Lectures at the US Army Command and General Staff College*, Combined Arms Center, For Leavenworth, KS, 1980, lecture IV, p. 8.

Select bibliography

UNPUBLISHED DOCUMENTS

Rowell to Vasey, 30 August 1942, Australian War Memorial: AWM 54, 225/2/5

Millotat, Christian, 'The Reinforced Panzergrenadier Battalion in Defence', Presentation to the Combat Training Centre Tactics Study Session, CFB Gagetown, 5–7 March 1984

OFFICIAL SOURCES

Association of the United States Army, *Women in Combat. Report to the President: The Presidential Commission on the Assignment of Women in the Armed Forces,* Brassey's, Washington, DC, 1993

Australian Army, *Land Warfare Doctrine, 1: The Fundamentals of Land Warfare,* Doctrine Wing, CATDC, Puckapunyal, VA 1999

Land Commander Australia, *Training Directive,* Sydney, 1998 (restricted)

Marine Corps Combat Development Command, *War Fighting Concepts for the 21st Century,* USMC, Quantico, VA, 1998

The Senate Standing Committee on Foreign Affairs, Defence and Trade, *Facing the Future Together: Report on Sexual Harassment in the Australian Defence Force,* Australian Government Publishing Service, Canberra, December 1994

United States General Accounting Office, National Security and International Affairs Division, *Report to the Honorable John W. Warner, U.S. Senate: Homosexuals in the Military Policies and Practices of Foreign Countries*, Washington, June 1993
United States General Accounting Office Report to the Secretary of Defense, *Women in the Military: Deployment in the Persian Gulf War*, Washington, July 1993

BOOKS AND MONOGRAPHS

Addis, Elisabetta, Russo, Valeria E. & Sebesta, Lorenza, (eds) *Women Soldiers: Images and Realities*, St Martin's Press, New York, 1994
Ahrenfeldt, R.H., *Psychiatry in the British Army in the Second World War*, Routledge & Kegan Paul, London, 1958
Allnutt, M.F., Haslam, D.R., Rejman, M.H. & Green, S., *Sustained Performance and Some Effects on the Design and Operation of Complex Systems*, Army Personnel Research Establishment, Ministry of Defence, London, 1990
Aurelius, Marcus, *Meditations,* Wordsworth, Hertfordshire, 1997 edn
Alvis, Michael W., *Dying for Peace: Understanding the Role of Casualties in US Peace Operations*, US Institute of Peace, Washington, DC, 1998 (microfiche)
Artwohl, A. & Christian, L., *Deadly Force Encounters*, Paladin Press, Boulder, CO, 1997
Bark, Dennis L. & Gress, David R., *A History of West Germany, Vol. 1: From Shadow to Substance, 1945–1963*, 2nd edn, Blackwell, Oxford, 1993
Bean, C.E.W., *The Official History of Australia in the War of 1914–18, Vol. III: The AIF in France*, Angus & Robertson, Sydney, 1942
Beaufre, General André, *An Introduction to Strategy*, Faber, London, 1965
Belenkey, G. (ed.), *Contemporary Studies in Combat Psychiatry*, Greenwood Press, London, 1987
Belenky, G., Balkin, T., Krueger, G., Headley, D. & Solick, R., *Effects of Continuous Operations (CONOPS) on Soldier and Unit Performance, Phase 1: Review of the Literature*, US Army Combined Arms Combat Development Activity, Fort Leavenworth, 1986
Belenky, G., Balkin, T., Krueger, G., Headley, D. & Solick, R., *Effects of Continuous Operations (CONOPS) on Soldier and Unit Performance: Review of the Literature and Strategies for Sustaining the Soldier in CONOPS*, Walter Reed Army Institute of Research, Bethesda, MD, 1987
Binkin, Martin & Eitelberg, Mark J. with Schexnider, Alvin J. & Smith, Marvin M., *Blacks and the Military*, Brookings Institution, Washington, DC, 1982
Binkin, Martin, *Who Will Fight the Next War?*, Brookings Institution, Washington, DC, 1993
Botting, Douglas, *In the Ruins of the Reich*, Grafton Books, London, 1985

Bourke, Joanna, *An Intimate History of Killing: Face-to-Face Killing in the Twentieth Century*, Granta Books, London, 1999

Breakwell, Glynis & Spacie, Keith, *Pressures Facing Commanders*, The Occasional no. 29, Strategic & Combat Studies Institute, Camberley, 1997

Breen, Bob, *A Little Bit of Hope, Australian Force—Somalia*, Allen & Unwin, Sydney, 1998

Burns, J., *The Brown and Blue Diamond at War*, 2/27th Battalion Ex-Servicemen's Association, Adelaide, 1960

Buscombe, Edward (ed.), *The BFI Companion to the Western*, André Deutsch, London, 1988

Catton, Bruce, *Grant Takes Command*, J.M. Dent, London, 1970

Cheeseman, Graeme & Hall, Robert A., *Preparing for Australia's Military After Next: The Price Report and a 'New Model' Australian Defence Force Academy*, Australian Defence Studies Centre, Canberra, 1997

Clausewitz, Carl von, *On War*, Michael Howard & Peter Paret (ed. & trans.), Princeton University Press, Princeton, NJ, 1989

Cocks, Doug, *People Policy: Australia's Population Choices*, University of New South Wales Press, Sydney, 1996

Cohen, Eliot A., *Citizens and Soldiers: The Dilemmas of Military Service*, Cornell University Press, Ithaca, NY, 1985

Connell, John, *Wavell: Scholar and Soldier*, Collins, London, 1964

Coulthard-Clark, Chris (ed.) *The Diggers: Makers of the Australian Military Tradition*, Melbourne University Press, Melbourne, 1993

Converse III, Elliott V. (ed.) *Forging the Sword: Selecting, Educating, and Training Cadets and Junior Officers in the Modern World*, Military History Symposium Series of the United States Air Force Academy, Vol. 5, Imprint Publications, Chicago, IL, 1998

Cooper, Cary L., *The Stress Check*, Prentice-Hall, Englewood Cliffs, NJ, 1981

Cox, Tom, *Stress*, Macmillan, London, 1978

Creveld, Martin van, *Technology and War: From 2000 BC to the Present*, Brassey's, London, 1991

Dear, I.C.B. & Foot, M.R.D. (eds), *The Oxford Companion to the Second World War*, Oxford University Press, Oxford, 1995

Deitchman, S.J. *Quantifying the Military Value of Training for System and Force Acquisition Decisions: An Appreciation of the State of the Art*, IDA Paper P–2881, Institute for Defense Analysis, Alexandria, VA, 1993

Dennis, Peter, Grey, Peter, Morris, Ewan, Prior, Robin with Connor, John (eds), *The Oxford Companion to Australian Military History*, Oxford, Melbourne, 1995

De Saxe, Maurice, *My Reveries upon the Art of War, Vol. 1: Roots of Strategy*, Stackpole Books, Harrisburg, PA, 1985

Douglas, Kirk, *The Ragman's Son: An Autobiography*, Pan Books, London, 1988

Downes, Cathy, *Social, Economic and Political Influences upon the Australian Army of the 1990s*, Strategic and Defence Studies Centre, Research School of Pacific Studies, Australian National University, Canberra, 1988

Driskell, J.E. & Salas, E. (eds), *Stress and Human Performance*, Lawrence Erlbaum Associates, Mahwah, NJ, 1996

du Picq, Ardant, *Battle Studies: Ancient and Modern Battle*, J.N. Greely & R.C. Cotton (trans.), Stackpole Books, Harrisburg, PA, 1987

Duffy, Christopher, *Red Storm on the Reich: The Soviet March on Germany, 1945*, Atheneum, New York, 1991

Dunlap, Jr, Charles J., *Technology and the 21st Century Battlefield: Recomplicating Moral Life for the Statesman and the Soldier*, Strategic Studies Institute, US Army War College, Carlisle, PA (15 January 1999)

Dyer, Gwynne, *War*, Crown Publishers, New York, 1985

Eichelberger, Robert L., *Our Jungle Road to Tokyo*, Battery Classics, Nashville, TN, 1989

Eddy, P. & Linklater, M. (eds), *The Falklands War*, Sunday Times Insight Team and Sphere Books, London, 1982

European Security Study, *Strengthening Conventional Deterrence in Europe: Proposals for the 1980s*, Macmillan, London, 1983

Flanagan, Stephen J., *NATO's Conventional Defences: Options for the Central Region*, Macmillan, London, 1988

Fletcher, Ian (ed.), *The Peninsular War: Aspects of the Struggle in the Iberian Peninsula*, Spellmount, Staplehurst, Kent, 1998

Flew, Anthony, *Thinking about Thinking (Or, Do I Sincerely Want to be Right?)*, Fontana, London, 1985

Freedman, Lawrence, *The Revolution in Strategic Affairs*, Adelphi Paper 318, Oxford University Press, Oxford, 1998

Fuller, J.F.C. *The Foundations of the Science of War*, Hutchinson, London, 1925

——*Generalship, Its Diseases and Their Cure*, Faber, London, 1933

Fussell, Paul, *The Boy Scout Handbook and Other Observations*, Oxford University Press, New York, 1982

——*Wartime*, Oxford University Press, New York, 1989

Gabriel, R.A., *No More Heroes: Madness and Psychiatry in War*, Hill & Wang, New York, 1987

Gal, R. & Mangelsdorff, A.D. (eds), *Handbook of Military Psychology*, John Wiley, New York, 1991

Garrard, John & Garrard, Carol (eds), *World War 2 and the Soviet People: Selected Papers from the Fourth World Congress for Soviet and East European Studies, Harrogate, 1990*, St Martin's Press, London, 1993

Gerster, Robin, *Big-Noting: The Heroic Theme in Australian War Writing*, Melbourne University Press, Melbourne, 1987

Glad, B. (ed.), *Psychological Dimensions of War*, Sage, Newbury Park, CA, 1990

Glass, D.C. & Singer, J.E., *Urban Stress*, Academic Press, New York, 1972

Gorman, P.F., *The Military Value of Training*, IDA Paper P–2515, Institute for Defense Analysis, Alexandria, VA, 1990

Gossett, Sue, *The Films and Career of Audie Murphy*, Empire Publishing, Madison, NC, 1996

Graham, Don, *No Name on the Bullet: A Biography of Audie Murphy*, Viking, New York, 1989

Grant, Ian, *Jacka, VC: Australia's Finest Fighting Soldier*, Macmillan, Melbourne, 1989

Gray, J. Glenn, *The Warriors: Reflections on Men in Battle*, Harper & Row, New York, 1959

Griffith, P. & Elmar, Dinter, *Not Over by Christmas*, Antony Bird, Chichester, 1983

Griffith, P., *Battle Tactics of the Civil War*, Yale University Press, New Haven, CT, 1989

——*The Ultimate Weaponry*, Sidgwick & Jackson, London, 1991

——*Battle Tactics of the Western Front: The British Art of Attack, 1916–18*, Yale University Press, New Haven, CT, 1994

Grossman, D., *On Killing*, Little Brown, London, 1996

Gullett, Henry 'Jo', *Good Company*, University of Queensland Press, Brisbane, 1992

Hallmark, B.W. & Crowley, J.C., *Company Performance at the National Training Center*, Rand, Washington, DC, 1997

Hamilton, Sir Ian, *A Staff Officer's Scrap Book during the Russo-Japanese War*, 2 vols, E. Arnold, London, 1906

Hancock, Eleanor, *The National Socialist Leadership and Total War, 1941–5*, St Martin's Press, New York, 1991

Hanson, Victor Davis & Heath, John, *Who Killed Homer? The Demise of Classical Education and the Recovery of Greek Wisdom*, The Free Press, New York, 1998

Harris, J.P. & Toase, F.N. (eds), *Armoured Warfare*, Batsford, London, 1990

Haslam, D.R., Allnutt, M.F., Worsley, D.E., Dunn, D., Abraham, J., Few, J., Labuc, S. & Lawrence, D.J., *The Effect of Continuous Operations upon the Military Performance of the Infantryman (Exercise Early Call)*, Army Personnel Research Establishment, Report 2/77, Farnborough, 1977

Herek, Gregory M., Jobe, Jared B. & Carney, Ralph M. (eds), *Out in Force: Sexual Orientation and the Military*, University of Chicago Press, Chicago, IL, 1996

Hetherington, John, *Blamey: Controversial Soldier*, Australian War Memorial, Canberra, 1973

Hiro, Dilip, *The Longest War: The Iran–Iraq Military Conflict*, Paladin, London, 1990

Hockey, G.R. (ed.), *Stress and Fatigue in Human Performance*, Wiley, New York, 1983

Holmes, Richard, *Firing Line*, Jonathan Cape, London, 1985

——*Acts of War: The Behavior of Men in Battle*, The Free Press, New York, 1985

Holz, R.F., Hiller, J.H., McFann, J.H. & H.H. (eds), *Determinants of Effective Unit Performance*, US Army Research Institute, Alexandria, VA, 1994

Horner, David M., *Australian Higher Command in the Vietnam War*, Strategic and Defence Studies Centre, Canberra, 1986

——*Blamey: The Commander-in-Chief*, Allen & Unwin, Sydney, 1998

Horowitz, S., Orlansky, J., Tillson, T.C., Gemeles, T.C., Gilman, H.J., Hammon, C. & Hoyler, H.M., *Unit Training in the Gulf War*, IDA Paper P–3087, Institute for Defense Analysis, Alexandria, VA, 1995

Hosking, Geoffrey, *A History of the Soviet Union 1917–1991*, final edn, Fontana Press, London, 1992

Howard, Michael, Andreopoulos, George J. & Shulman, Mark R. (eds), *The Laws of War: Constraints on Warfare in the Western World*, Yale University Press, New Haven, CT, 1994

Howe, Renate (ed.), *Women and the State: Australian Perspectives* (special edition of *Journal of Australian Studies*) no. 37, Bundoora, 1993

Jones, David E, *Women Warriors: A History*, Brassey's, Washington, DC, 1997

Jones, James, *The Thin Red Line*, Charles Scribner's Sons, New York, 1951

Karpen, Ulrich (ed.), *The Constitution of the Federal Republic of Germany: Essays on the Basic Rights and Principles of the Basic Law with a Translation of the Basic Law*, Nomos Verlagsgesellschaft, Baden-Baden, 1988

Keegan, John, *The Face of Battle*, Jonathan Cape, London, 1976

——*A History of Warfare*, Hutchinson, London, 1993

Keegan, J. & Holmes, R., *Soldiers*, Hamish Hamilton, London, 1985

Kenney, George C., *General Kenney Reports*, Office of Air Force History Reprint, Washington, DC, 1987

Knightley, Philip, *The First Casualty*, Harcourt Brace Jovanovich, New York, 1975

Kovic, Ron, *Born on the Fourth of July*, McGraw-Hill, New York, 1976

Larson, Eric V., *Casualties and Consensus: The Historical Role of Casualties in Domestic Support for U.S. Military Operations*, Rand, Santa Monica, CA, 1996

Leonhard, Robert R., *The Art of Maneuver: Maneuver-Warfare Theory and AirLand Battle*, Presidio, Novato, CA, 1991

Liddell Hart, B.H., *Thoughts on War*, Faber, London, 1943

Lind, William S., *Maneuver Warfare Handbook*, Westview Press, Boulder, CO, 1985

Linderman, Gerald F., *Embattled Courage: The Experience of Combat in the American Civil War,* The Free Press, New York, 1987

——*The World Within War: America's Combat Experience in World War II*, The Free Press, New York, 1997

Livi-Bacci, Massimo, *A Concise History of World Population*, Blackwell, Cambridge, MA, 1992

Lorber, Judith, *Paradoxes of Gender*, Yale University Press, New Haven, CT, 1994

Lorentzen, Lois Ann & Turpin, Jennifer (eds), *The Women and War Reader*, New York University Press, New York, 1998

Lucas, T.W, Bankes, S.C. & Vye, P., *Improving the Analytic Contribution of Advanced Warfighting Experiment*, Rand Documented Briefing DB–207-A, Washington, DC, 1998

Mandelbaum, Michael, 'Learning to be warless', *Survival*, vol. 41, no. 2 (Summer) 1999

Marshall, S.L.A., *Men Against Fire*, Infantry Journal Press, Washington, DC, 1947

——*The Soldier's Load and the Mobility of a Nation*, Marine Corps Association, Quantico, VA, 1987 (first published 1950)

McAulay, Lex, *The Battle of Long Tan: 'The Legend of ANZAC Upheld'*, Arrow edn, London, 1987

——*The Battle of Coral*, Arrow, Sydney, 1988

McKercher, B.J.C. & Hennessy, Michael (eds), *The Operational Art: Developments in the Theories of War*, Praeger, Westport, CT, 1996

McNab, Andy, *Bravo Two Zero*, Corgi Books, London, 1997

McNamara, Robert S., *In Retrospect: The Tragedy and Lessons of Vietnam*, Times Books, New York, 1995

McNeill, Ian, *To Long Tan*, Allen & Unwin, Sydney, 1993

Meichenbaum, D., *Stress Inoculation Training*, Pergamon, New York, 1985

Mitchell, Brian, *Women in the Military: Flirting with Disaster*, Regnery, Washington, DC, 1998

Mitchell, Brian, *Weak Link: The Feminization of the American Military*, Regnery Gateway, Washington, DC, 1989

Montgomery, B.L., *A History of Warfare*, Collins, London, 1968

Moskos, Charles C. & Butler, John Sibley, *All That We Can Be: Black Leadership and Racial Integration the Army Way*, Basic Books, New York, 1996

Moskos, Charles C. & Chambers II, John Whiteclay (eds), *The New Conscientious Objection: From Sacred to Secular Resistance*, Oxford University Press, Oxford, 1993

Mueller, John E., *War, Presidents and Public Opinion*, Wiley, New York, 1973

——*Policy and Opinion in the Gulf War*, Chicago University Press, Chicago, 1994

Muir, Kate, *Arms and the Woman*, Coronet Books, London, 1993

Murphy, Audie, *To Hell and Back*, 8th edn, Holt, Rinehart & Winston, New York, 1971

O'Brien, Tim, *The Things They Carried*, Harper Collins, New York, 1990

O'Hanlon, Michael, 'Can high technology bring US troops home?', *Foreign Policy*, no. 113 (Winter) 1998–99

Ofer, Dalia & Weitzman, Lenore J. (eds), *Women in the Holocaust*, Yale University Press, New Haven, CT, 1998

Overy, Richard, *Russia's War: Blood Upon the Snow*, TV Books, New York, 1997

Patrick, James, *A Glasgow Gang Observed*, Eyre Methuen, London, 1973

Pearson, Patricia, *When She Was Bad: How Women Get Away with Murder*, Virago, London, 1998

Pew, R.W. & Mavor, A.S. (eds), *Representing Human Behavior in Military Simulations: Interim Report*, National Academy Press, Washington, DC, 1997

Phillips, T.R. (ed.), *Roots of Strategy*, Military Service Publishing, Harrisburg, PA, 1943

Pierre, Andrew J. (ed.), *The Conventional Defense of Europe: New Technologies and New Strategies*, Council on Foreign Relations, New York, 1986

Pimlott, John & Badsey, Stephen (eds), *The Gulf War Assessed*, Arms & Armour, London, 1992

Preston, Richard A., Roland, Alex & Wise, Sydney F., *Men in Arms: A History of Warfare and its Interrelationships with Western Society*, 5th edn, Harcourt Brace Jovanovich College, Fort Worth, TX, 1990

Pyle, Ernie, *Here is Your War*, Lancer Books, New York, 1943

Radford, Jill & Russell, Diana E.H. (eds), *Femicide: The Politics of Woman Killing*, Twayne Publishers, New York, 1992

Read, Anthony & Fisher, David, *The Fall of Berlin*, Pimlico, London, 1993

Reid, Brian Holden & Dewar, Michael (eds), *Military Strategy in a Changing Europe*, Brassey's, London, 1991

Robertson, John, *Anzac and Empire*, Hamlyn, Melbourne, 1990

Rogan, Helen, *Mixed Company: Women in the Modern Army*, Putnam's, New York, 1981

Rooney, D., Kallmeier, V. & Stevens, G., *Mission Command and Battlefield Digitization: Human Sciences Considerations*, Report no. DERA/CHS/ HS3/ CR980097/1.0, DERA, Centre for Human Sciences, Farnborough, 1998

Ross, Jane, *The Myth of the Digger: The Australian Soldier in Two World Wars*, Hale & Iremonger, Sydney, 1985

Rowland, D., 'The effect of combat degradation on the urban battle', *Journal of the Operational Research Society*, vol. 42, no. 7, 1991, pp. 543–53

Rule, E.J., *Jacka's Mob*, Angus & Robertson, Sydney, 1933

Sarkesian, Sam C. (ed.), *Combat Effectiveness: Cohesion, Stress and the Volunteer Military*, Sage, Beverley Hills, CA, 1980

Schwarz, Benjamin C., *Casualties, Public Opinion, and US Military Intervention*, Rand, Santa Monica, CA, 1994

Schwarzkopf, General Norman H., *It Doesn't Take a Hero*, Linda Grey Bantam Books, New York, 1992

Scott, Wilbur J. & Stanley, Sandra Carson (eds), *Gays and Lesbians in the Military: Issues, Concerns and Contrasts*, Aldine de Gruyter, New York, 1994

Senger und Etterlin, General Frido, *Neither Fear nor Hope*, George Malcolm (trans.), Macdonald, London, 1960

Shaw, Martin, *Post-Military Society: Militarism, Demilitarization and War at the End of the Twentieth Century*, Polity Press, Cambridge, 1991

Siddle, B.K., *Sharpening the Warrior's Edge: The Psychology and Science of Training*, PPCT Management Systems, Millstadt, IL, 1995

Simpkin, Richard E., *Race to the Swift: Thoughts on Twenty-First Century Warfare*, Brassey's, London, 1985

Simpson, Harold B. *Audie Murphy: American Soldier*, Hill College Press, Hillsboro, TX, 1982

——*Simpson Speaks on History*, Hill College Press, Hillsboro, TX, 1986

Simpson, H., *Evaluating Large-Scale Training Simulations*, Defense Manpower Data Center, Washington, DC, 1998

Sorley, Lewis, *Honorable Warrior: General Harold K. Johnson and the Ethics of Command*, University Press of Kansas, Lawrence, KS, 1998

Sperling, Gotz, *German Perspectives on the Defence of Europe: An Analysis of Alternate Approaches to NATO Strategy*, Queen's Centre for International Relations, Kingston, Ontario, 1985

Spiller, Roger J. (ed.), *S.L.A. Marshall at Leavenworth: Five Lectures at the US Army Command and General Staff College*, Combined Arms Center, Fort Leavenworth, KS, 1980

Stebbing, L. Susan, *Thinking to Some Purpose*, Penguin, Harmondsworth, 1959

Stiehm, Judith Hicks, *Bring Me Men and Women: Mandated Change at the US Air Force Academy*, University of California Press, Berkeley, CA, 1981

Stouffer, S.A., *The American Soldier: Combat and its Aftermath, Studies in Social Psychology in World War II, Vol. 2*, Princeton University Press, Princeton, NJ, 1949

Terry, Ray, *Women in Khaki: The Story of the British Woman Soldier*, Columbus, London, 1988

Thompson, T.J., Pleban, R.J. & Valentine, P.J., *Determinants of Effective Unit Performance: Battle Staff Training and Synchronization*, US Army Research Institute for the Behavioral and Social Sciences, Alexandria, VA, 1994

Van Creveld, Martin, *The Transformation of War*, Free Press, New York, 1991

Wavell, Sir Archibald, *Generals and Generalship*, Times Publishing, London, 1941

Weigley, Russell, *The Age of Battles*, Pimlico, London, 1993

Weiner, E.L., Kanki, B.G. & Helmreich, R.L. (eds), *Crew Resource Management*, Academic Press, New York, 1993

Welford, A.T., *Fundamentals of Skill*, Methuen, London, 1967

Wheelwright, Julie, *Amazons and Military Maids: Women who Dressed as Men in the Pursuit of Life, Liberty and Happiness*, Pandora, London, 1989

Wills, Garry, *John Wayne: The Politics of Celebrity*, Faber & Faber, London, 1997

Winter, Denis, *Death's Men: Soldiers of the Great War*, Allen Lane, London, 1978

Zsambok, C.E. & Klein, G. (eds), *Naturalistic Decision Making*, Lawrence Earlbaum, Mahwah, NJ, 1997

Zweig, Paul, *The Adventurer: The Fate of Adventure in the Western World*, Princeton University Press, Princeton, NJ, 1974

ARTICLES

Appel, J.W. & Beebe, G.W., 'Preventive psychiatry', *Journal of the American Medical Association*, vol. 131, no. 18, 1946, pp. 1469–76

Berkun, M.M., 'Performance decrement under psychological stress', *Human Factors*, vol. 6, 1964, pp. 21–30

Berkun, M.M., Bialek, H.M., Kern, R.P. & Yagi, K., 'Experimental studies of psychological stress in man', *Psychological Monographs*, vol. 9, no. 534, 1962

Brinkworth, Lisa, 'Twisted Sisters', *Harpers & Queen*, April 1997, pp. 131–3

Cohen, Eliot A., 'Tocqueville on war', *Social Philosophy & Policy*, vol. 3, no. 1 (Autumn) 1985

Coker, Christopher, 'Post-modern war', *RUSI Journal*, vol. 143, no. 3, June 1998

Cooper, Lieutenant General Charles G., 'The day it became the longest war', *Proceedings of the US Naval Institute*, May 1996

Cott, Nancy F., 'Marriage and women's citizenship in the United States, 1830–1934', *The American Historical Review*, vol. 103, no. 5, December 1998, pp. 1440–74.

'Diagnosis Female', *Time*, 8 March 1999, p. 46

Dobell, Graeme, 'The Media's Perspective on Peacekeeping', in Hugh Smith (ed.), *Peacekeeping: Challenges for the Future*, Australian Defence Studies Centre, Canberra, 1993

Dobie, Kathy, 'A few nice men', *US Vogue*, February 1993, p. 299

Downes, Cathy, 'An ethos for an army in the twenty-first century', in Hugh Smith (ed.), *Preparing Future Leaders*, Australian Defence Studies Centre, Canberra, 1998

Driskell, J.E., Willis, R.P. & Cooper, C., 'Effect of overlearning on retention', *Journal of Applied Psychology*, vol. 77, 1992, pp. 615–22

Ehrenreich, Barbara, 'The real truth about the female', *Time*, 8 March 1999, p. 38

Eikenberry, Karl W., 'Take no casualties', *Parameters*, vol. 26, no. 2 (Summer) 1996

Endsley, M.R., 'Towards a theory of situation awareness in dynamic systems', *Human Factors*, vol. 37, no. 1, 1995, pp. 32–64

Findlay, Trevor, *Cambodia: The Legacy and Lessons of UNTAC*, SIPRI Research Report no. 9, Oxford University Press, 1995

Flach, J.M., 'Situation awareness: proceed with caution', *Human Factors*, vol. 37, no. 1, 1995, pp. 149–57

Frank, G.R. 'Continuous operations', *ARMOR*, March/April, 1982, pp. 19–23

Friedland, N. & Keinan, G., 'Training effective performance in stressful situations: three approaches and implications for combat training', *Military Psychology*, vol. 4, 1992, p. 159

Fukuyama, Francis, 'Women and the evolution of world politics', *Foreign Affairs*, vol. 77, no. 5, September/October 1998

Gal-or, Y., Tenenbaum, G., Furst, G. & Shertzer, M., 'Effect of self-control and anxiety on training performance in young and novice parachuters', *Perceptual and Motor Skills*, vol. 60, 1985, pp. 743–6

Gartner, Scott Sigmund & Segura, Gary M., 'War, casualties, and public opinion', *Journal of Conflict Resolution*, vol. 42, no. 3 (June) 1998

Geiselman, R.E. & Samet, M.G. 'Summarizing military information: an application of schema theory', *Human Factors*, vol. 22, no. 6, 1980, pp. 693–705

Gessert, Robert A., 'The Airland battle and NATO's new doctrinal debate', *RUSI: Journal of the Royal United Services Institute for Defence Studies*, June 1984

Griffith, J., 'The army's new unit personnel replacement and its relationship to unit cohesion and social support', *Military Psychology*, vol. 1, 1989, pp. 17–34.

Griffith, P., 'Countering surprise by mobility, a concept for armoured warfare on the central front', *The Sandhurst Journal*, vol. 1, no 1, 1990

Haney, C., Banks, W.C. & Zimbardo, P.G., 'Interpersonal dynamics in a simulated prison', *International Journal of Criminology and Penology*, vol. 1, 1973, pp. 69–97

Hanley, Lynne, 'To kill and be killed', *The Women's Review of Books*, vol. 8, no. 6, March 1991, p. 3

Harig, Paul T., 'The digital general: reflections on leadership in the post-information age', *Parameters*, Autumn 1996, p. 139

Inglis, K.S., 'Anzac and the Australian military tradition', Revue Internationale d'Histoire Militaire, Canberra, No. 72, 1990

International Institute for Strategic Studies, 'The problem of combat reluctance', *Strategic Survey 1995/96*, Oxford University Press, Oxford, pp. 48–57

Kemnitz, Heidemarie, 'Mädchen und Militär: "Weiblichkeit" im Diskurs um nationale Frauendienste, Wehrerziehung für Mädchen und Frauen in der Bundeswehr', *Feministische Studien*, 16/1, 1998

Kull, Steven, 'What the public knows that Washington doesn't', *Foreign Policy*, no. 101 (Winter) 1995/96

Lewis-Young, Peter, 'Australia's Project Wundurra', *Asian Defence Journal*, no. 7, 1997, p. 20.

——'Australia's Future Infantryman', *Armed Forces Journal International*, February 1999, p. 44.

Luttwak, Edward N., 'Where are the great powers? At home with the kids', *Foreign Affairs*, vol. 73, no. 4 (July/August) 1994)

——'Toward post-heroic warfare', *Foreign Affairs*, vol. 74, no. 3 (May/June) 1995

——'A post-heroic military policy', *Foreign Affairs*, vol. 75, no. 4 (July/ August) 1996

——'From Vietnam to *Desert Fox*: civil–military relations in modern democracies', *Survival*, vol. 41, no. 1 (Spring) 1999)

Malone, David, 'Haiti and the international community: a case study', *Survival*, vol. 39, no. 2 (Summer) 1997

McClusky, M.R., Fowlkes, J.E., Pierce, L.G. & Dwyer, D.J., 'Measurement of command/control staff performance in tactical training environments', Proceedings of the Interservice/Industry Training, Simulation and Education Conference, Orlando, FA, 1998

Mermin, Jonathan, 'Television news and American intervention in Somalia: the myth of a media-driven foreign policy', *Political Science Quarterly*, vol. 112, no. 3 (Fall) 1997

Milburn, T.S., 'Casualties—the crucial factor in modern conflicts', *The British Army Review*, no. 113 (August) 1996

Milgram, S., 'Behavioral study of obedience', *Journal of Abnormal and Social Psychology*, vol. 67, 1963, pp. 371–8

Morgan, Robin, 'Complicated trust', *Ms Magazine*, January/February 1991, p. 1

Morley, Morris & McGillion, Chris, '"Disobedient" generals and the politics of redemocratization: the Clinton administration and Haiti', *Political Science Quarterly*, vol. 112, no. 3 (Fall) 1997

Moskos, Charles C., 'When Americans accept casualties', *Newsletter*, Inter-University Seminar on Armed Forces and Society (Winter) 1996

Mueller, John E. 'Changing attitudes towards war: the impact of the First World War', *British Journal of Political Science*, vol. 21, no. 2 (January) 1991

Nogami, Glenda Y., Brander, Christopher D. & Slusser, Karen A., '21st century strategic leader competencies: from the army generals' perspective', Proceedings of the 39th Annual Conference of the International Military Testing Association, Sydney, 1997, p. 340

Norris, F.H. & Murrell, S.A., 'Prior experience as a moderator of disaster impact on anxiety symptoms in older adults', *American Journal of Community Psychology*, vol. 16, 1998, pp. 665–83

Rogers, General Bernard W., 'Follow-on forces attack (FOFA): myths and realities', *NATO Review*, no. 6, December 1984, pp. 1–9

Roush, E. 'Combat exclusion: military necessity or another name for bigotry?', *MINERVA: Quarterly Report on Women and the Military*, vol. 8, no. 3, 1990

Rowland, D., 'The effect of combat degradation on the urban battle', *Journal of the Operational Research Society*, vol. 42, no. 7, 1991, pp. 543–53

Sarpolsky, Harvey M. & Shapiro, Jeremy, 'Casualties, technology, and America's future wars', *Parameters* vol. 26, no. 2 (Summer) 1996

Saunders, T., Driskell, J.E., Johnston, J.H. & Salas, E., 'The effect of stress inoculation training on anxiety and performance', *Journal of Occupational Health Psychology*, vol. 1, 1996, pp. 170–86

Slim, Field Marshal Sir William, 'Higher command in war', *Military Review*, May 1990, pp. 10–21

Smith, Hugh & McAllister, Ian, 'The changing military profession: integrating women in the Australian Defence Force', *Australia and New Zealand Journal of Sociology*, vol. 27, no. 3 (November) 1991

Spiller, Roger J., 'Isen's run: human dimensions of warfare in the 20th century', *Military Review*, May 1988, LXIX, v

——'Shell shock', *American Heritage*, May/June 1988

——'The tenth imperative', *Military Review*, April 1989, LXIX, iv

——'My guns: a memoir of the Second World War', *American Heritage*, December 1991

——'The Price of Valor', *Military History Quarterly*, Spring 1993, V, iii

——'War in the dark', *American Heritage*, February/March 1999

Stein, Janice Gross, 'Deterrence and compellence in the Gulf, 1990–91', *International Security*, vol. 17, no. 2 (Fall) 1992

Steiner, M. & Neumann, M., 'Traumatic neurosis and social support in the Yom Kippur War returnees', *Military Medicine*, vol. 143, 1978, pp. 866–8

Stevenson, Charles A., 'The evolving Clinton doctrine on the use of force', *Armed Forces & Society*, vol. 22, no. 4 (Summer) 1996

Sutton, Boyd D., Landry, John R. et al, 'New directions in conventional defence?', *Survival* (March/April) 1984, pp. 50–70

Swank, R.L. & Marchand, W.E. 'Combat neuroses: development of combat exhaustion', *Archives of Neurology and Psychology*, vol. 55, 1946

Thayer, Carlyle A., 'Vietnam: a critical analysis', in Peter R. Young (ed.), *Defence and the Media in Time of Limited War*, Frank Cass, London, 1992

Timmerman, Frederick W., 'Human dimensions of the battlefield', *Military Review*, April 1989, LXIX, iv

Tziner, A. & Vardi, Y., 'Effect of command style and group cohesiveness on the performance effectiveness of self-selected tank crews', *Journal of Applied Psychology*, vol. 67, 1982, pp. 769–75

Welford, A.T., 'Mental work-load as a function of demand, capacity, strategy and skill', *Ergonomics*, vol. 21, no. 3, 1978, pp. 151–67

Wertkin, R.A., 'Stress-inoculation training: principles and applications', *The Journal of Contemporary Social Work*, vol. 66, 1985, pp. 611–16.

Zajonc, R.B., 'Social facilitation', *Science*, vol. 149, 1965, pp. 269–74

Zengel, Patricia, 'Assassination and the law of armed conflict', *Military Law Review*, vol. 134 (Fall) 1991)

NEWSPAPERS

Audie Murphy Research Foundation, *Newsletters*, Vols 1–8, 1997–99

Graham, Bob, 'Right stuff wrong sex', *The Australian Magazine*, 23–24 January 1999, pp. 11–17

Greene, Gervase, 'Military women in danger of being a passing parade', *The Age*, 15 January 1999, p. 1

——'Crisis among defence force women denied', *The Age*, 21 January 1999, p. 5

Gunn, Michelle, 'Boom or doom', *The Australian*, 2 April 1999, p. 15

'Hillary Clinton blasts Taliban', *The Age*, 6 March 1999, p. 20

Jacobson, Philip, 'Blood sisters', *The Australian* Magazine, March 2–3 1996, pp. 12–18

Maddison, Sarah, 'Right to fight is feminism's frontline', *The Weekend Australian*, 9–10 January 1999, p. 21

Nafisi, Azar, 'The veiled threat in Iran', *The Australian*, 6–7 March 1999

Pogonelska, Paulyne, 'Females in a fury', *The Age*, 4 February 1999, p. 16

Scarborough, Rowan, 'Study: Annapolis women's knees more injury prone', *The Washington Times*, 12 July 1998

ELECTRONIC SOURCES

'Bring justice to victims of war-time rape in Bosnia', *http://www.now.org/issues/global/bosnia.html* (accessed 5 March 1999)

'C. Frauen in den Streitkräften anderer Staaten', 'Frauen in der Bundeswehr', *http://www.bundeswehr.de/bundeswehr/frauen/c-anderek.htm* (accessed 9 March 1999)

Chief Land Staff, *Gender Integration Study: The Experience of Women Who Have Served in the Combat Arms*, *http://www.dnd.ca/eng/min/reports/cls-gis/sepsum.htm* (accessed 5 March 1999)

Corbett, Dawn, 'Women around the globe face threats to human rights', *http://www.now.org/nnt/fall–98/global.html* (accessed 5 March 1999)

Powell, General Colin, speech of 10 November 1998, *http://www.mod.uk/policy/equalops/poweladd.htm* (accessed 5 March 1999)

'Stop the abuse of women and girls in Afghanistan!', *http://www.now.org/foundation/global/taliban.html* (accessed 5 March 1999)

'Strength of the defence forces', *http://www.mil.f/english/s20.html* (accessed 9 March 1999)

'The defence of Sweden', *http://www.mil.se/intro/intro4_e.html* (accessed 9 March 1999)

The Department of National Defence (Canada) homepage, *http://www.dnd.ca/*

The Kassebaum Baker report on gender-integrated training: *http://www.defenselink.mil/pubs/g.t* (accessed 9 March 1999)

'Vision 2010—the integrated Navy', 1998, *http://www.dnd.ca/navy/marcom/tin/summ_e.htm* (accessed 5 March 1999)

Whittaker, Michael, 'We are not alone—women in NATO', Defence 2000, *http://131.137.255.5/vcds/mcert/cnrs/dec98/art08_e.asp* (accessed 5 March 1999)

'Women in the Canadian Forces', February 1998, *http://www.dnd.ca/eng/archive/may98/women_b_e.htm* (accessed 5 March 1999)

FILM

Jacka VC, A film by Nigel Buesst & Roger Cooper, Monash University, 1977

To Hell and Back, Universal Pictures, California, 1955

Index